# Springer Tracts in Civil Engineering

**Series Editors**

Sheng-Hong Chen, School of Water Resources and Hydropower Engineering, Wuhan University, Wuhan, China

Marco di Prisco, Politecnico di Milano, Milano, Italy

Ioannis Vayas, Institute of Steel Structures, National Technical University of Athens, Athens, Greece

**Springer Tracts in Civil Engineering** (STCE) publishes the latest developments in Civil Engineering - quickly, informally and in top quality. The series scope includes monographs, professional books, graduate textbooks and edited volumes, as well as outstanding PhD theses. Its goal is to cover all the main branches of civil engineering, both theoretical and applied, including:

- Construction and Structural Mechanics
- Building Materials
- Concrete, Steel and Timber Structures
- Geotechnical Engineering
- Earthquake Engineering
- Coastal Engineering; Ocean and Offshore Engineering
- Hydraulics, Hydrology and Water Resources Engineering
- Environmental Engineering and Sustainability
- Structural Health and Monitoring
- Surveying and Geographical Information Systems
- Heating, Ventilation and Air Conditioning (HVAC)
- Transportation and Traffic
- Risk Analysis
- Safety and Security

**Indexed by Scopus**

To submit a proposal or request further information, please contact:
Pierpaolo Riva at Pierpaolo.Riva@springer.com (Europe and Americas) Wayne Hu at wayne.hu@springer.com (China)

More information about this series at https://link.springer.com/bookseries/15088

Marc Panet · Jean Sulem

# Convergence-Confinement Method for Tunnel Design

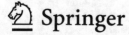 Springer

Marc Panet
Académie des Technologies
Paris, France

Jean Sulem
Laboratoire Navier
École des Ponts ParisTech
Marne-la-Vallée, France

ISSN 2366-259X          ISSN 2366-2603  (electronic)
Springer Tracts in Civil Engineering
ISBN 978-3-030-93195-7          ISBN 978-3-030-93193-3  (eBook)
https://doi.org/10.1007/978-3-030-93193-3

Translation from the French language edition: *Le calcul des tunnels par la méthode convergence-confinement* by Marc Panet, and Jean Sulem, © Authors 2021. Published by Presses des Ponts. All Rights Reserved.

This Springer imprint is published by the registered company Springer Nature Switzerland AG
The registered company address is: Gewerbestrasse 11, 6330 Cham, Switzerland

# Foreword

This book is a translation into English of *Le calcul des tunnels par la méthode convergence-confinement*, which is the second edition of a book with the same title written by Marc Panet in 1995. The new edition is authored by Marc Panet and Jean Sulem and reflects the intervening developments in tunnel construction, design, and analysis.

The convergence-confinement method, often called the characteristic curve method, was originally simply a concept to understand ground–structure interaction. Starting in the 1970s and particularly in the two books, it has been combined with analytical and empirical formulations to make it into a rigorous design approach. The underlying principle is the formal relation of the behavior of the ground in the axial direction of the tunnel with the ground–structure interaction in the transverse cross section. The analytical and empirical developments of the method make it now possible to include a variety of effects that are important in tunneling such as different construction methods, water, and time dependence.

Reflecting what just has been said, the book is much more than an introduction to the convergence-confinement method. It starts by introducing the reader to different construction methods and different tunnel supports, in the first two chapters. This is then followed by chapters on elastic, elastoplastic, and time-dependent behavior as well as one on numerical methods. While written in the context of the convergence-confinement method, these chapters can also be used independently to familiarize the reader with the particular topic.

To summarize: The book not only brings the convergence-confinement method to the present state of the art and practice, but it can serve as a valuable handbook for any reader interested in and working in tunnel design.

Herbert Einstein
Professor of Civil and Environmental
Engineering
Massachusetts Institute of Technology
Cambridge, MA, USA

# Contents

# Chapter 1
# Introduction

## 1.1 Construction Methods

The design of a tunnel includes the choice of an excavation process and of the type of support and lining. Along the last years, the tunnel construction methods have greatly changed.

### 1.1.1 Conventional Construction Methods

In the conventional methods, the ground excavation is done by picks, shovels or road headers in soils or soft rocks. It is carried out by drilled and blast in hard rocks (Fig. 1.1). A distinction is to be done between the support installed close to the heading and the final lining formerly in masonry and usually in cast concrete.

During the nineteenth century and most of the twentieth century, full-face excavation was performed only for small section galleries; the excavation of large tunnels was made by dividing the cross section. Different types of cross section division were used and named according to the European country where they were originated: Belgian, Deutsch, Austrian etc. A semi section driving is commonly used for large sections (Fig. 1.2).

In the past, the support was made by wooden struts. Support by steel arches, bolts and shotcrete brought important changes in the construction methods.

The full-face excavation of large tunnels was applied only in grounds of good quality. New techniques of presupport and of stabilization of the face led Pietro Lunardi (Lunardi 2008) to develop the ADECO-RS method with full-face excavation, bolting of the face, presupport and final concrete lining with a floor cast in place close to the heading (Fig. 1.3). Many examples, especially in Italy, proved the ADECO-RS method to be efficient both technically and economically in difficult grounds.

© The Author(s), under exclusive license to Springer Nature Switzerland AG 2022
M. Panet and J. Sulem, *Convergence-Confinement Method for Tunnel Design*,
Springer Tracts in Civil Engineering, https://doi.org/10.1007/978-3-030-93193-3_1

**Fig. 1.1** Drilling with a jumbo for a full-face blasting (Atlas Copco)

**Fig. 1.2** Tunnel driven in semi section on the motorway A86 (France)

**Fig. 1.3** South-east high-speed train in France. Tartaiguille tunnel. Application of the ADECO-RS method with fiber glass bolting of the face

## 1.1.2 Tunnel Boring Machines (TBM)

Tunnel boring machines are more and more common because they permit the largest advancement rates. In good conditions, the advancement rate may be over 20 m per day. However, driving incidents often due to difficult ground conditions or machine breakdowns may cause long stops and increase the delay of construction.

Three dates may be considered as landmarks of the TBM development:

- 1888: Two exploratory galleries were dug in chalk on each Channel shore by Colonel Beaumont.
- 1952: James Robbins built the first modern TBM for the driving of a hydroelectric gallery of Oake Dam in South Dakota
- 1964: The first TBM with a confining of the face by air pressure was built by the Robbins Company for the construction of the underground metro RER A in Paris between Place de l'Etoile and La Défense.

The types of TBM may be distinguished according to:

- The mode of excavation: road header or full face cutterhead.
- The presence and the type of shield: no shield, single or double shield.

**Fig. 1.4** Double shield TBM (Herrenknecht)

- The way to push forward the machine by grippers leaning on the walls, by thrust rams acting against segment lining installed at the tail skin.
- The confinement or not of the working face.
- The mode of mucking (Fig. 1.4).

To stabilize the working face or to limit the ground deformations, the face may be confined. The confinement may be achieved by compressed air, earth pressure or slurry (Fig. 1.5).

In compressed air TBM, the air pressure may be difficult to maintain. Muck is extracted by a pressure relief system. In some highly permeable grounds, air leakage may induce instabilities at the face.

In the earth pressure balance machines (EPBM), the earth pressure in the cutterhead chamber is maintained by controlling the extraction of the muck by a screw conveyor. The pressure confinement acting on the face is not homogeneous, and it is difficult, even impossible, to keep the cutterhead chamber full of material.

The use of a bentonite slurry in the cutterhead chamber is undoubtedly the most secure confinement method of the face. The efficiency is due to the formation of a cake at the surface of the ground face. The muck mixed with the slurry is flushed out by a fluid system which controls the discharge rate to maintain the confining pressure. The need of a treatment plant to separate the slurry from the muck increases the cost of this technique.

TBM manufacturers are continuously innovating to adapt the TBM to various ground conditions. The mixed face TBM may function in open and closed mode with different confining and mucking techniques.

**Fig. 1.5 a** Slurry pressure TBM (Herrenknecht). **b** Earth Pressure Balance Machine (Herrenknecht). **c** Mixed face TBM (Herrenknecht)

## 1.2 Displacements Induced by Tunnel Construction

### 1.2.1 Tunnel Convergence

The modern tunnel design methods rely on the analysis of the ground displacement field around the excavation. The ground and support displacements are monitored as the face advances.

Convergence measurements are now widespread in underground works. The origin of the convergence measurements is in the mining work sites with the measurement of the relative displacement of the floor and the roof of the mined layer behind the heading.

In a section orthogonal to the axis of a tunnel the convergence in the direction $\alpha$ is the relative displacement of two opposite points of the section in this direction:

$$C_\alpha = -\Delta L_\alpha \tag{1.1}$$

where $L_\alpha$ is the initial distance between the two points. The convergence is said to be positive if the distance shortens and vice versa for a negative convergence.

The convergence is measured by mechanical or optical distance-meters with a relative accuracy of about $10^{-5}$. Optical distance-meters are more and more used.

In a given section the convergence measured in several directions underlines the anisotropy of the deformation.

Even without theoretical analysis, the monitoring of the convergences proved to be extremely useful to control the ground behavior and to check the adequacy of the support. It gives essential data for any observational method.

In each section the convergences increase with the distance to the face. If no instability occurs, the convergence gradient diminishes as the heading progresses. One must try to install the convergence targets as close as possible from the face. It is not always possible, for instance because of the presence of a shield.

At a given time, the positions of the face and of the monitored section are given by their coordinates $X_f$ and $X$ respectively along the tunnel axis. The distance of the considered section from the face is:

$$x = X_f - X \tag{1.2}$$

The curve $x(t)$ describes the advance of the face and is necessary for the interpretation of the curve of convergence.

If there is no time-dependent deformation, the convergence is only dependent on $x$. At a distance $d_f$ called the distance of influence of the face, the face is far enough from the considered section, and the convergence does not increase any more.

If time dependent deformations occur, the convergence continues to increase with time beyond the distance of influence of the face. The analysis of the convergence must consider the curve $x(t)$ which describes the advance of the face, the curve

$C(x)$ which describes the convergence as a function of the distance from the face and the curve $C(t)$ which describes the convergence as a function of time (Fig. 1.6). Considering these curves for successive monitored sections and the periods of stop of

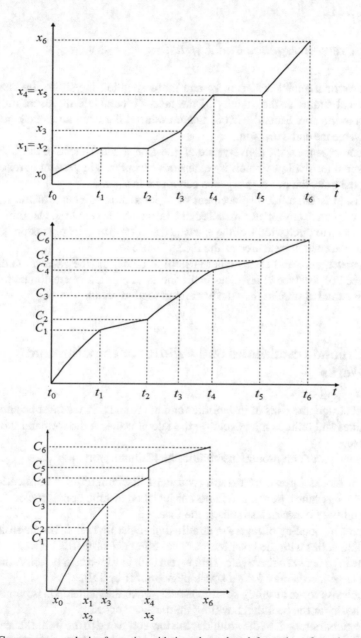

**Fig. 1.6** Convergence analysis of a section with time-dependent deformations: face advance versus time—convergence versus face advance—convergence versus time

advance of the face, the distance of influence of the face may be determined. These periods corresponding to a stop of the face advance also give useful information about the ground rheological behavior.

### 1.2.2   Pre-convergence and Extrusion

Nowadays for a tunnel design more and more attention is paid to the analysis of stresses and strains in the vicinity of the face. To complete the usual monitoring of the convergence, Lunardi (2008) has introduced the complementary notions of pre-convergence and extrusion.

Pre-convergence is the convergence of a section situated ahead of the face. It may be measured for shallow tunnels by extensometers placed in vertical boreholes. But most of the time, the pre-convergence cannot be measured.

Extrusion is the axial displacement of a point situated ahead of the face. It is measured by an extensometer installed in a horizontal borehole the length of which is usually two to three widths of the excavated section. The total extension is the sum of the displacements measured as the face advances.

Pre-convergence and extrusion are useful to determine the excavation disturbed zone ahead of the face. This disturbed zone is very often overestimated. In most cases, its extent is smaller than the excavated section width.

## 1.3   Ground Instabilities and Failures in Underground Works

In tunnel design, the risks of instabilities and of ground failures must be anticipated. Instabilities and failures may occur on the lateral walls, at the roof and the floor or on the face.

The more common ground instabilities and failures are:

- Fall of blocks formed by natural rock mass discontinuities (stratification beds, schistosity planes, joints, and faults) or included in soils or soft rocks.
- Instability of cohesionless soils at the face.
- Failures by bending of layers in stratified or schistose rocks. The intensity of the bedding varies with the layer inertia. The difference of inertia between beds may create gaps between the layers. Thin layers with low inertia may fail. A thick layer with large inertia may form a stable plane roof (Fig. 1.7).
- In schistose rocks or thinly stratified rocks, when the tunnel axis is parallel to the schistosity or the bedding, buckling failures may occur.
- If the global strength of the ground is too low compared to the initial state of stress, the development of shear or extension fractures creates a damaged zone around the excavation. In hard and brittle rocks, spalling at the walls of the excavation may

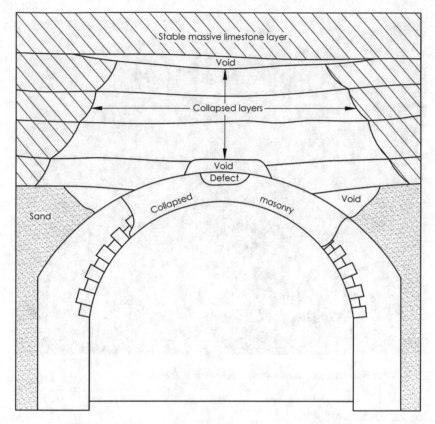

**Fig. 1.7** Vierzy tunnel (France). Failure on 16 June 1972. Bending and failure of the limestone layers of Lower Lutetian. Stability of the thick Saint Leu bed

give brutal failures called rock bursts (Fig. 1.8). In more ductile rocks, squeezing phenomena occur with the development of large convergences (Fig. 1.9).

The most severe instabilities in tunneling construction which in the past caused a great number of casualties are due to sudden large water and debris inflows at the heading. The blow out of the face under high water pressure may happen when crossing faults with cohesionless infillings, mylonized zones, or karstic and glacial channels (Fig. 1.10). Hydrogeological conditions of the ground must be carefully investigated and analyzed. They are not studied in this book.

## 1.4  Support Functions

The support of a tunnel has the following functions:

**Fig. 1.8** Mont Blanc tunnel (France-Italy). Rockbursts at the walls

- To secure the work site.
- To control the ground deformations which should remain compatible with the tunnel's functionality and that of the neighboring structures.
- To accommodate the ground pressure.

In a conventional excavation method, the support is installed as the heading progresses. It may be set in a single or in several phases. It may include a presupport ahead of the face. The final support consists most of the time of a concrete lining which must resist to the long-term ground pressures.

With a TBM equipped with a shield, the support behind the face is brought by the shield. Behind the shield, the immediate and final supports are secured by a ring made of precast segments. The ring of segments is installed under the skin of the shield. The gap between the ground and the outer face of the ring is filled by grouting a mortar.

The various types of support and their mechanical characteristics are described in detail in Chap. 2.

**Fig. 1.9**  Lyon-Turin tunnel. Saint-Martin-La-Porte access gallery. Large convergences in schistose rocks (courtesy of Eiffage)

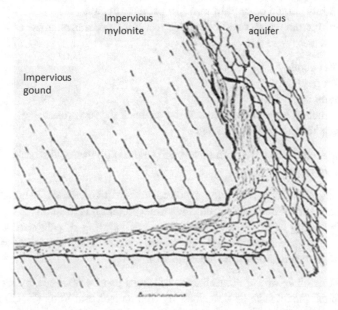

**Fig. 1.10**  Water and debris inflows due to the blow out of the face under high water pressure (C.Bordet)

## 1.5    Time-Dependent Deformation

The convergences which go on increasing beyond the distance of influence of the face are due the time-dependent behavior of the ground.

The two causes of differed deformations are the rheological behavior of the ground which exhibits creep-relaxation phenomena and the evolution of water pore pressures around the excavation, the tunnel acting as a drain if the support and the lining are not watertight. Even when the lining is watertight, most often a drainage system is installed at the outer face of the lining. When the ground has a low permeability, the time to reach the steady state flow regime may be long.

Among the causes of differed deformations in underground works, the case of the swelling rocks, especially in clay and gypsum rocks requires special attention. Tunneling in swelling ground is not addressed in this book.

## 1.6    Natural State of Stress

The excavation of a tunnel perturbs the static equilibrium of the ground. The mechanical analysis of the excavation requires the knowledge of the initial state of stress. But the determination of the initial state of stress in a rock mass, especially at great depth, is a challenging task and most of the time there is a great uncertainty about it. This uncertainty should be kept in mind in projects design.

The main factors which must be considered for the determination of the natural state of stress are:

- The tunnel depth.
- The site topography.
- The ground behavior.
- The presence of large shear zones at the scale of the rock mass.
- The regional tectonic state of stress.

Different techniques have been developed to measure the local stresses. There are mainly of two types:

- Overcoring tests can be done only in sound and unfractured rock. In heterogeneous rocks the results are scattered, and their interpretation is difficult.
- Hydrofracturing of the rock mass and hydrofracturing of preexisting fractures (HTPF) may give good results but in most cases, they are insufficient for a complete determination of the stress tensor.

The best estimate of the natural state of stress can be obtained by a numerical geomechanical model such as described by Stephanson and Zang (2012) including all available data of the main parameters.

For shallow tunnels driven under a horizontal ground surface, it is usually assumed that the vertical stress $\sigma_v^0$ is a principal stress given by the weight of the overlying ground. The principal horizontal stresses $\sigma_h^0$ are given by:

$$\sigma_h^0 = K_0 \sigma_v^0 \tag{1.3}$$

$K_0$ coefficient in a semi-infinite elastic mass is equal to: $K_0 = \frac{\nu}{1-\nu}$ where $\nu$ is the Poisson ratio.

In particular conditions as in overconsolidated soils, the coefficient $K_0$ may be larger than 1.

## 1.7  Choice Tunnel Support and Lining

The natural and troglodyte caves prove that underground structures may remain stable on a very long period without any support. However, most of the recent tunnels are supported and lined because they are dug in grounds in which instabilities may occur in the short or long term.

To design the support and the lining of an underground structure, the engineers may choose between empirical methods mainly based on experience and analytical methods more in adequacy with their usual training in structural design.

However, new methods of excavation and new techniques were created by pragmatic engineers having a fair understanding of the ground behavior in underground structures. Theoretical explanations more or less adequate were brought later after successful achievements met on the work sites. Such was the case of the New Austrian Method with a support made mainly of shotcrete and bolts, a method now widely used all over the world.

Two types of support design methods are to be considered:

- Empirical methods mainly based on geotechnical classifications for tunnels.
- More recent methods resulting from the analysis of the ground support interaction and using numerical models.

### 1.7.1  Geotechnical Classifications for Tunnels

Empirical design consists to reproduce the design which has proved to be efficient in previous works with more or less similar conditions. Similar conditions refer to depth, geological formations, section of the tunnel. The engineer reproduces the excavation and support techniques used previously and introduces recent improvements.

The ground behavior may be inferred from the geological conditions and from the tunnel depth. The main types of ground behavior are:

- Stable grounds with small convergences,
- Cohesionless soil or heavily fractured rocks, unstable at the heading without water flow,

- Cohesionless soils or heavily fractured rocks with water flow. A drainage and/or a ground treatment by grouting or freezing is necessary at the heading,
- Sound rock masses with structurally controlled blocks fall,
- Grounds with a stable face and moderate convergences,
- Grounds with an unstable face and large convergences (squeezing rocks),
- Rock masses with brittle failure on the walls of the excavation (spalling) with the possibility of rockbursts,
- Grounds with time dependent deformations,
- Swelling rocks.

These behaviors are not exclusive one from the other. Grounds of various behavior may be present in the same section.

From these types of behavior, the main ground parameters to consider in the design are:

- The ground nature: cohesionless soils, cohesive soils, rocks with plastic/ductile behavior, rocks with brittle behavior, rocks with viscous behavior, swelling rocks;
- The structure of the rock mass: type and frequency of the discontinuities;
- The hydrogeological conditions: presence of a water table, ground permeability, water head;
- The strength characteristics of the soil or rock mass compared to the natural state of stress;
- The short-term and long-term deformation characteristics of the ground.

The geotechnical classifications for tunnels are based on some of these parameters. Because of the simplicity of their use, they are widely spread all over the world.

The three most common classifications (see appendix) are:

- The Rock Mass Rating of Z. T. Bieniawski (Bieniawski 1983, 1993).
- The Q coefficient of Nick Barton (Barton et al. 1974).
- The Geological Strength Index (GSI) of Evert Hoek (Hoek 1994).

The coefficient given by these classifications permits to determine directly the mode of excavation and the type of support. Correlations have been established between the coefficients RMR, Q and GSI.

The GSI classification also permits to evaluate the parameters of an elastoplastic law: deformation moduli, yield criterion. In many sites where the results of laboratory and in situ tests are scattered, the GSI approach is interesting to estimate the mechanical parameters at the tunnel scale. However, the given values must be analyzed with care considering all the geological and geotechnical investigations, comparing the available data with those collected on similar sites. The GSI approach may also be useful for the analysis of the ground support interaction.

## 1.7.2 Design Methods Based on the Analysis of the Ground Support Interaction

Compared to classification methods, a quantitative analysis of the ground-support interaction is more complex. It is necessary to know the natural state of stress, the short-term and long-term constitutive behavior of the ground, the parameters of the constitutive models of the ground and of the support, the phasing of the excavation and of the support installation. Furthermore, a tridimensional analysis is required to take into account the face advance. All these features may now be included in numerical models, but the great number of parameters and the uncertainty of many of them increase the complexity of these models. The interpretation of the results is difficult without many computations to test the sensitivity of the results to the uncertain data. Three-dimensional numerical models are still rarely used at the project stage and are mainly used for a posteriori interpretation of observations and measurements made during the excavation.

The convergence-confinement method replaces the complex three-dimensional analysis with a simplified two-dimensional analysis. The face advance is simulated by a decrease of the stresses acting on the wall of the tunnel. This method is didactic for an easy understanding of the ground-support interaction but is based on strong assumptions whose relevance is to be checked for each application.

## 1.8 Principles of the Convergence-Confinement Method

In the convergence-confinement method (Panet 1976, 1993), the complexity of the tridimensional problem near the face is reduced to a plane strain problem by applying a stress state at the wall of the tunnel defined as:

$$\sigma = (1 - \lambda)\sigma_0 \tag{1.4}$$

where $\sigma_0$ is the natural state of stress.

The coefficient $\lambda$ is called the *deconfinement ratio* or *stress release ratio*. It is equal to 0 in the initial state and increases as the face advances to reach 1 in the final state if the tunnel is unsupported (Fig. 1.11). This coefficient is derived from the *Longitudinal Displacement Profile* (LDP), which is the curve giving the evolution of the radial displacement $u$ at the tunnel wall as a function of the distance to the face.

The relationship between the stress and the radial displacement at the wall of the tunnel:

$$f_m(\sigma, u) = 0 \tag{1.5}$$

gives the ground convergence, represented by the *Ground Reaction Curve* (GRC).

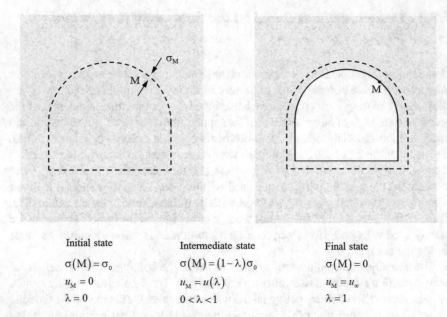

|                      |                           |                  |
|----------------------|---------------------------|------------------|
| Initial state        | Intermediate state        | Final state      |
| $\sigma(M) = \sigma_0$ | $\sigma(M) = (1-\lambda)\sigma_0$ | $\sigma(M) = 0$ |
| $u_M = 0$            | $u_M = u(\lambda)$        | $u_M = u_\infty$ |
| $\lambda = 0$        | $0 < \lambda < 1$         | $\lambda = 1$    |

**Fig. 1.11** The progressive deconfinement of the ground with the face advance

The mechanical behavior of the support is described by the relationship between the stress at the outer face of the support and the radial displacement:

$$f_s(\sigma, u) = 0 \tag{1.6}$$

The support which limits the convergence is set at a distance $d$ from the face, called the unsupported distance, where the radial displacement is $u_d$. The relationship:

$$f_s(\sigma, u - u_d) = 0 \tag{1.7}$$

characterizes the response of the support. It is the confinement law of the support represented by the *Support Confining Curve* (SCC).

The final static equilibrium of the ground-support interaction is given by the solution of the system of the two equations, the ground convergence law and the support confinement law.

In the case of axi-symmetry, it is possible to give a very simple graphical representation of the convergence-confinement method. The final equilibrium is given by the intersection of the GRC and of the SCC (Fig. 1.12).

In the literature, the GRC and the SCC are often called the characteristic curves of the ground and of the support respectively. They were first introduced by Pacher (1964) for a qualitative analysis of the ground-support interaction. The introduction of the deconfinement ratio in order to account for the displacement of the wall at the

**Fig. 1.12** Determination of
the ground pressure at final
equilibrium in the
axisymmetric case

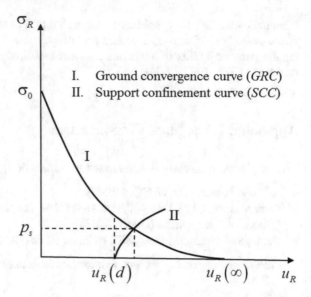

I.  Ground convergence curve (*GRC*)
II. Support confinement curve (*SCC*)

unsupported distance from the face opened the way for a quantitative analysis (Panet and Guellec 1974).

In the convergence-confinement method, the ground is considered as a homogeneous medium without discontinuities. The behavior of the ground is often complex; the choice of its constitutive law implies a simplification of the real behavior and a homogenization at the scale of the work. The determination of the parameters of the constitutive law of the ground is a difficult task; they cannot be simply drawn from laboratory and/or in situ tests and must result from the analysis of all qualitative and quantitative data of the geological and geotechnical investigations and considering their scatters, their uncertainties and their limits.

Soils and rocks are complex materials. The development of numerical models permits to introduce constitutive laws which are more and more complex with a great number of parameters. These possibilities may seduce; but they are often limited to academic works and difficult to apply in practical cases. A simple model whose shortcomings and limitations can be analyzed may be preferred to a more sophisticated one with a great number of parameters whose evaluation is uncertain and which would require a large number of computations to assess the sensitivity of the results to large uncertainties. The expertise acquired by the experience and the common sense is the best tool to solve complex problems.

The convergence-confinement method is not only useful for the design, but may be used during the excavation of the tunnel to interpret the data of the monitoring in the observational method.

The fundamentals of the convergence-confinement method are developed analytically with restrictive assumptions: tunnel with a circular section, a full face excavation, a homogeneous and isotropic ground, an isotropic state of natural stress. This formulation has not only a didactic interest but allows to have a quick and approximate evaluation of convergence and ground pressure, to test the sensitivity of the results to the variability of the parameters and to validate at least partially numerical models (AFTES 1983, 2002).

## Appendix: Rock Mass Classifications

The main rock mass classifications used for tunnel design are:

- AFTES classification (AFTES 2003).
- Rock Mass Rating (RMR) of Bieniawski (Bieniawski 1993).
- Q index of Nick Barton (Barton 2013).
- Geological Strength Index (GSI) of Evert Hoek (Hoek 1994).

  Each classification has its own features, its advantages and drawbacks.

### *AFTES Classification*

The classification of the *Association Française des Travaux en Souterrain* (AFTES) gives a well-ordered quantitative description of the main characteristics of rock masses for tunnel design (AFTES, 2003):

- The rock characteristics determined in laboratory at the sample scale.
- The various discontinuity indexes of the rock mass and the nature and the characteristics of the discontinuity sets.
- The rock mass characteristics at the scale of the structure: mechanical and hydrogeological parameters, natural state of stress.

  AFTES classification gives useful guidelines for the geological and geotechnical investigations and for the establishment of the contract between the project owner and the contractors.
  This classification is only a reference for the design engineer but does not give any guideline for the choice of the type of the excavation process and of the type of support. Therefore, it differs for the three following classifications.

### *Rock Mass Rating*

Several ground classifications were proposed for tunneling in the past. The most known are Karl Terzaghi's in English speaking countries, Lauffer's in German speaking countries, Protodiakonov's in Eastern European countries. Z. T. Bieniawski proposed in 1973 the first classification based on a new discipline, rock mechanics. It introduced improvements in the description and the behavior of rock masses.

The parameters in the last issue of the RMR classification (1989) are scores related to various ground conditions (Table 1.1):

- $R1$: Strength of the rock given by uniaxial compression tests or the Franklin index.
- $R2$: Rock Quality Designation (RQD) introduced by Deere (1963), (see also Deere and Deere 1988) defined in a cored borehole as the percentage of intact cores longer than 100 mm.
- $R3$: Discontinuities spacing.
- $R4$: Conditions of the discontinuity surfaces.
- $R5$: Groundwater conditions according to the water inflows in the tunnel.

$A$ is a correction parameter which depends on the dip and direction versus the axis of the tunnel of the main discontinuities.

The coefficient $RMR$ is obtained by:

$$RMR = R1 + R2 + R2 + R3 + R4 + R5 + A$$

This simple additive rule has no theoretical basis.

The $RMR$ coefficient takes values between 0 and 100.

It may be noticed that this classification emphasizes the discontinuous state of the rock mass with the three parameters $R2$, $R3$ and $R4$. In heterogeneous rocks, the choice of $R1$ is difficult. If no cored borehole has been drilled, no data are available to determine $R2$.

Guidelines for the excavation and the support of the tunnel have been established according to the $RMR$ coefficient and the unsupported span of the section.


## Q Quality Index

The Q index was developed by Nick Barton between 1971 and 1974 at the Norwegian Geotechnical Institute. Later it has been improved taking account for the development of new tunneling techniques (Barton et al. 1974, Barton 2013).

The parameters included in the Q index are:

- The $RQD$ index.
- The joint set number $J_n$.
- The joint roughness number $J_r$ of the major discontinuities.
- The joint alteration number $J_a$ characterizing the weathering of the surfaces or the infilling of the discontinuities.
- The joint water reduction $J_w$.
- The stress reduction factor $SRF$ which take account of the weakness zones, of the rock strength compared to the natural state of stress, of the possibilities of squeezing or swelling of the rock mass.

**Table 1.1** *RMR* classification

*(a) Classification parameters and their ratings*

| Parameter | | Range of values//rating | | | | |
|---|---|---|---|---|---|---|
| 1 | Uniaxial compressive strength (MPa) | >250 | 100–250 | 50–100 | 25–50 | 5–25 1–5 <1 |
| | **Rating R1** | **15** | **12** | **7** | **4** | **2 1 0** |
| 2 | Drill core quality RQD (%) | 90–100 | 75–90 | 50–75 | 25–50 | <25 |
| | **Rating R2** | **20** | **17** | **13** | **8** | **5** |
| 3 | Spacing of discontinuities | >2 m | 0.6–2 m | 20–60 cm | 6–20 cm | <6 cm |
| | **Rating R3** | **20** | **15** | **10** | **8** | **5** |
| 4 | Condition of discontinuities | • Very rough<br>• No separation<br>• Non persistent<br>• Unweathered wall rock | • Rough<br>• Aperture <0.1 mm<br>• Non persistent<br>• Slightly weathered wall rock | • Slightly rough<br>• Aperture <0.1 mm<br>• Persistent<br>• Highly weathered wall rock | • Smooth<br>• Aperture 1–5 mm<br>• Persistent<br>• Infilling <5 mm | • Aperture >5 mm<br>• Persistent<br>• Infilling >5 mm |
| | **Rating R4** | **30** | **25** | **20** | **10** | **0** |
| 5 | Groundwater | Inflow per 10 m tunnel length (L/min) | none | <10 | 10–25 | 25–125 | >125 |
| | | General conditions | completely dry | damp | wet | Dripping | Flowing |
| | **Rating R5** | **15** | **10** | **7** | **4** | **0** |

(continued)

**Table 1.1** (continued)

*(b) Effect of discontinuity strike and dip orientation in tunneling (A)*

| Strike perpendicular to tunnel axis | | | | Strike parallel to tunnel axis | | Irrespective of strike |
|---|---|---|---|---|---|---|
| Drive with dip | | Drive against dip | | | | |
| Dip 45°–90° | Dip 20°–45° | Dip 45°–90° | Dip 20°–45° | Dip 45°–90° | Dip 20°–45° | Dip 0°–20° |
| Very favourable | Favourable | Fair | Unfavourable | Very unfavourable | Fair | Fair |
| 0 | –2 | –5 | –10 | –12 | –5 | –5 |

*(c) Total rating and rock mass classes*

| Rock mass rating | 100–81 | 80–61 | 60–41 | 40–21 | <20 |
|---|---|---|---|---|---|
| Class | I | II | III | IV | V |
| Description | Very good rock | Good rock | Fair rock | Poor rock | Very poor rock |

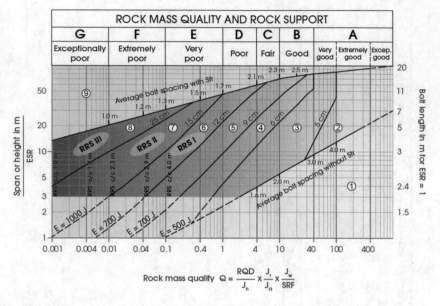

Fig. 1.13 The Q-system chart for the design of rock support. From Barton (2013)

- The value of the $Q$ index is given by:

$$Q = \frac{RQD}{J_n} \times \frac{J_r}{J_a} \times \frac{J_w}{SRF}$$

The rock mass discontinuities and their characteristics play a major role in the definition of the $Q$ index.

The $SRF$ is another important factor which may have large variations, between 1 and 400. Four types of rock mass are considered in the classification.

As the RMR, the Q index is the result of the analysis of a great number of empirical data and case studies.

Nick Barton established a chart for the design support based on the Q value and the Excavation Support Ratio (ESR) (Fig. 1.13).

## Ground Strength Index (GSI)

A new classification was introduced in 1994 by Evert Hoek (Hoek 1994) considering the difficulty to describe correctly rock masses by scalars. It is based on a qualitative description of the structure of rock masses and of the surface conditions of discontinuities.

The average GSI is given by Table 1.2.

**Table 1.2** Geological strength index for a discontinuous rock mass. From Hoek and Marinos (2000)

| GEOLOGICAL STRENGTH INDEX FOR JOINTED ROCKS | SURFACE CONDITIONS | | | | |
|---|---|---|---|---|---|
| | VERY GOOD | GOOD | FAIR | POOR | VERY POOR |
| STRUCTURE | DECREASING SURFACE QUALITY ⟶ | | | | |
| INTACT OR MASSIVE–intact rock cpecimens or masslve in situ rock with few widely spaced discontinuilities | 90<br>80 | | | | |
| BLOCKY–well interlocked un-disturbed rock mass consistion of cubical blocks formed by three intersecting discontinuity sets | | 70<br>60 | | | |
| VERY BLOCKY–interlocked, partially disturbed mass with multi-faceted angular blocks formed by 4 or more joint sets | | | | | |
| BLOCKY/DISTURBED/SEAMY –folded with angular blocks formed by many intersecting discontinuity sets. Persistence of beding planes or schistosity | | | 40 | | |
| DISINTERATED–poorly inter-locked, heavily broken rock mass with mixture of angular and rounded rock pieces | | | | 20 | |
| LAMINATED/SHEARED–Lack of bockiness due to close spacing of weak schistosity or shear planes | | | | | 10 |

*(Left axis: DECREASING INTERLOCKING OF ROCK PIECES ↓)*

The GSI is comprised between 0 and 100. A very accurate value of the GSI has no meaning. The uncertainty of the GSI value is about 10–20.

A correlation between the RMR and the GSI has been proposed:

$$GSI = RMR - 5 \quad \text{with } R5 = 15$$

As the other classifications, its application to heterogeneous rock masses is hazardous. However, Hoek and Marinos (2000) have extended the GSI to hetero-geneous formations as the flysch met during the Athens Metro construction.

# References

AFTES (1983) Recommandations pour l'emploi de la méthode convergence-confinement. Tunnels et Ouvrages Souterrains 59:219–238

AFTES (2002) Recommandations relatives à la méthode convergence-confinement. Tunnels et Ouvrages Souterrains 170:79–89
AFTES (2003) Caractérisation des massifs rocheux utile à l'étude et à la réalisation des ouvrages souterrains. Tunnels et Espaces Souterrains 177:140–186
Barton N, Lien R, Lunde J (1974) Engineering classification of rock masses for the design of tunnel support. Rock Mech 6(4):189–236
Barton N (2013) Using the Q-system—rock mass classification and support design. Norwegian Geotechnical Institute, Oslo
Bieniawski ZT (1983) Rock mechanics design in mining and tunnelling. Balkema, Rotterdam
Bieniawski ZT (1993) Classification of rock masses for engineering: the RMR system and future trends. In: Comprehensive rock engineering, vol 3, 22, Pergamon Press, London, pp 553–573
Deere DU (1963) Technical description of rock cores for engineering purpose. Rock Mech Eng Geol 1(1):16–22
Deere DU, Deere DW (1988) The rock quality designation (RQD) index in practice. In: Kirkaldie L (ed) Rock Classification Systems for Engineering Purposes, ASTM STP 984. American Society for Testing and Materials, Philadelphia, pp 91–101
Hoek E (1994) Strength of rock and rock masses. ISRM News J 2(2):4–16
Hoek E, Marinos (2000). Predicting tunnel squeezing problems in weak heterogeneous rock masses. Tunnels Tunnel Int 32(11):45–51
Lunardi P (2008) Design and construction of tunnels. Springer, Berlin
Pacher F (1964) Deformationsmessungen im Versuchsstollen als Mittel zur Erforschung des Gebirgsverhaltens und zur Bemessung des Ausbaues. In: Müller L (ed) Grundfragen auf dem Gebiete der Geomechanik/principles in the field of geomechanics. Felsmechanik und Ingenieurgeologie/Rock Mechanics and Engineering Geology, vol 1. Springer, Berlin
Panet M (1976) Stabilité et soutènements des tunnels. La Mécanique des roches appliquée aux ouvrages du génie civil. Presses de l'Ecole Nationale des Ponts et Chaussées, Paris, pp 145–166
Panet M (1993) Understanding deformations in tunnels. In: Comprehensive rock engineering, vol 1, 27, Pergamon Press, London, pp 663–690
Panet M, Guellec P (1974) Contribution à l'étude du soutènement derrière le front de taille. In: Proceedings of 3rd congress of the International Society for Rock Mechanics, Denver, vol 2, part B, pp 1130–1134
Stephanson O, Zang A (2012) ISRM suggested methods for stress estimation, part.5: establishing a model for the in situ stress at a given site. Rock Mech Rock Eng 45(6):955–969

## Additional Reading

Bouvard-Lecoanet A, Colombet G, Esteule F (1988) Ouvrages souterrains—conception, Réalisation, Entretien. Presses de l'Ecole Nationale des Ponts et Chaussées, Paris
Comité Français de Mécanique des Roches (2013) Manuel de Mécanique des Roches. Tome IV. Part.7, Tunnels à grande profondeur. Collection Sciences de la Terre et de l'Environnement. Presses des Mines, Paris
Cundall PA, Hart RD (1993) Numerical modelling of discontinua. In: Comprehensive rock engineering, vol 2, 9, Pergamon Press, London, pp 231–234
Daemen JJK, Fairhurst C (1972) Rock failure and support loading. In: Proceedings of international symposium. Underground Openings, Lucerne, pp 356–369
Duffaut P (1977) Ground pressure and tunnelling from the nineteenth century to present. Underground Space 1(3):185–200
Goodman RE, Shi GH (1985) Block theory and its application to rock engineering. Prentice Hall, London

Guenot A, Panet M, Sulem J (1985) A new aspect in tunnel closure interpretation. In: Ashworth E (ed), Proceedings of 26th U.S. symposium on rock mechanics. Rapid City, South Dakota, USA. Balkema, pp 455–460

Hoek E, Brown ET (1980) Underground excavations in rocks. Institution of Mining and Metallurgy, London

Lombardi G (1973) Dimension of tunnel lining with regard to constructional procedure. Tunnels Tunnell 5:340–351

Mandel J (1959) Les calculs en matière de pressions des terrains. Revue de l'Industrie Minérale, pp 78–92

Mahtab MA, Grasso P (1992) Geomechanics principles in the design of tunnels and caverns in rocks. Devel Geotech Eng 72. Elsevier

Rabcewicz LV (1964–1965) The new Austrian tunneling method. Part I. Water Power, Nov 1964, 453–457, Part II Dec 1964, 511–515, part III, Jan 1965, 19–24

Rabcewicz LV, Golser J (1973) Principles of dimensioning the support system for the "New Austrian tunnelling method". Water Power, pp 88–93

Salençon J (1994) Pratique de la modélisation en géotechnique. In: Proceedings of international conference on soil mechanics and foundation engineering, New-Dehli, India, pp 261–262

Terzaghi K (1946) Rock defects and loads on tunnel supports. In: Rock tunnelling with steel supports rock defects and loads on tunnel supports, commercial shearing and stamping company, Youngstown, Ohio, pp 15–99

Ward WH (1978) Ground supports for tunnels in weak rocks. 18th Rankine lecture. Géotechnique 28:133–171

# Chapter 2
# Methods of Support

The temporary or final support of a tunnel may be achieved by a great variety of techniques. This chapter briefly describes the main methods of support. Its objective is to provide the necessary data to determine the support confinement curve SCC.

## 2.1 Unsupported Distance

In conventional methods without a presupport, the support is placed at some distance of the face. During excavation, the largest distance between the face and the support is called the unsupported distance or unsupported span $d$.

The unsupported distance is smaller than the distance of influence of the face to control the convergences and to avoid instabilities behind the face. It depends on the rate of advance of the face. If the advance of the face is stopped, it may be necessary to reduce, even to suppress, the unsupported distance.

Bieniawski (1989, 1993) proposed a chart giving the unsupported distance versus the stand-up time and the RMR coefficient (Fig. 2.1).

In the convergence- confinement method, a key parameter, the value of the deconfinement ratio $\lambda$ at the unsupported distance, is to be determined.

## 2.2 Normal Stiffness and Bending Stiffness of a Support

In its interaction with the ground, the mechanical behavior of the support may be characterized by the relationships between the stresses acting on the outer face of the support and its deformation.

In the simple case of axi-symmetry, in the domain of elastic behavior of the support, a single modulus $K_s$ characterizes the support stiffness (Fig. 2.2).

© The Author(s), under exclusive license to Springer Nature Switzerland AG 2022
M. Panet and J. Sulem, *Convergence-Confinement Method for Tunnel Design*,
Springer Tracts in Civil Engineering, https://doi.org/10.1007/978-3-030-93193-3_2

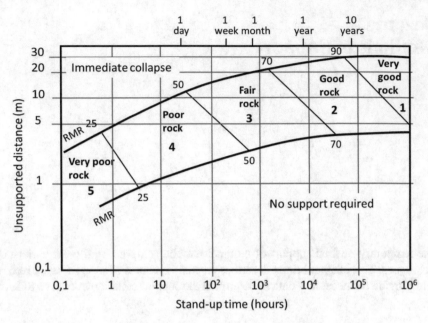

**Fig. 2.1** Bieniawski's chart of the unsupported distance versus the ground stand-up time (after Bieniawski 1989, 1993)

**Fig. 2.2** Definition of the stiffness of the support in the axisymmetric case

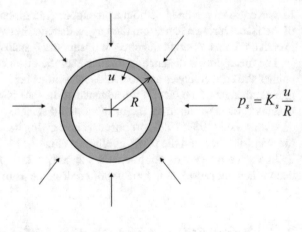

$$p_s = K_s \frac{u}{R} \qquad (2.1)$$

where

$p_s$   is the pressure exerted by the ground on the support,
$u$    is the uniform radial displacement of the support,
$R$    is the radius of the circular section.

In a more general case, the stiffness may be characterized by two moduli, the normal stiffness and the bending stiffness.

## 2.3 Normal Stiffness and Bending Stiffness of a Circular Cylindrical Shell

In many cases (cast concrete lining, precast segments, shotcrete support, steel sets), the support is modelled as an equivalent homogeneous thin cylindrical shell with a thickness $e$. The stiffness characteristics of the shell in elasticity are derived from the relationships between the stresses acting at the outer face of the shell and the corresponding displacements.

For a tunnel with a circular section of radius $R$ (Fig. 2.3), these equations are written as:

$$\sigma_R = K_{sn}\left(\frac{u}{R} + \frac{1}{R}\frac{dv}{d\theta}\right) + K_{sn}\left(\frac{1}{R}\frac{d^4u}{d\theta^4} + \frac{2}{R}\frac{d^2u}{d\theta^2} + \frac{u}{R}\right)$$

$$\tau_{R\theta} = -K_{sn}\left(\frac{1}{R}\frac{du}{d\theta} + \frac{1}{R}\frac{d^2v}{d\theta^2}\right) \tag{2.2}$$

where,

$\sigma_R$ and $\tau_{R\theta}$    are the radial stress and the shear stress at the outer face of the shell,
$u$               is the radial displacement,
$v$               is the hoop displacement.

The stiffness of the cylindrical shell is characterized by two moduli $K_{sn}$ and $K_{sf}$.

$K_{sn}$ is the normal stiffness,

$$K_{sn} = \frac{E_s}{1 - v_s^2}\frac{e}{R} \tag{2.3}$$

**Fig. 2.3** Stresses acting on the cylindrical shell

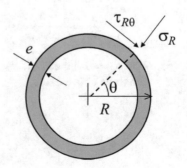

$K_{sf}$ is the bending stiffness,

$$K_{sf} = \frac{E_s}{1 - v_s^2} \frac{I}{R^3} \tag{2.4}$$

$E_{sf}$ and $v_s$ are the Young's modulus and the Poisson's ratio of the shell's material, $I = \frac{e^3}{12}$ is the modulus of inertia of the shell.

It may be noticed that:

$$\frac{K_{sn}}{K_{sf}} = 12\left(\frac{R}{e}\right)^2 \tag{2.5}$$

Thus, the bending stiffness modulus is much smaller than the normal stiffness modulus.

For $\frac{R}{e} = 10$, $K_{sn} = 1200K_{sf}$.

If the assumption of a thin shell is not fulfilled, the equations of a thick tube may be used. Then the normal stiffness modulus is given by:

$$K_{sn} = \frac{E_s\left[R^2 - (R - e)^2\right]}{(1 + v_s)\left[(1 - 2v_s)R^2 + (R - e)^2\right]} \tag{2.6}$$

Equations (2.1) can be easily integrated in the following simple loading cases:

(a)  $\sigma_R = p$,  $\tau_{R\theta} = 0$

$$\frac{u}{R} = \frac{1}{K_{sn} + K_{sf}}p, \quad v = 0 \tag{2.7}$$

(b)  $\sigma_R = q\cos 2\theta$,  $\tau_{R\theta} = 0$

$$\frac{u}{R} = \frac{1}{9K_{sf}}q\cos 2\theta, \quad \frac{v}{R} = -\frac{1}{18K_{sf}}q\sin 2\theta \tag{2.8}$$

(c)  $\sigma_R = 0$,  $\tau_{R\theta} = t\sin 2\theta$

$$\frac{u}{R} = -\frac{1}{18K_{sf}}t\sin 2\theta, \quad \frac{v}{R} = \frac{1}{4}\left[\frac{1}{K_{sn}} + \frac{1}{9K_{sf}}\right]t\sin 2\theta \tag{2.9}$$

(d)  For: $\sigma_R = p + q\cos 2\theta$, $\tau_{R\theta} = t\sin 2\theta$, the solution is obtained by superposing the three above cases:

$$\frac{u}{R} = \frac{1}{K_{sn} + K_{sf}}p + \frac{1}{9K_{sf}}\left[q - \frac{1}{2}t\right]\cos 2\theta$$

$$\frac{v}{R} = \frac{1}{4K_{sn}}t\sin 2\theta - \frac{1}{18K_{sf}}\left[q - \frac{1}{2}t\right]\sin 2\theta \qquad (2.10)$$

The normal force $N$ (positive in compression) and the bending moment $M$ are given by the equations:

$$\frac{1}{R}N + \frac{1}{R^2}\frac{d^2M}{d\theta^2} = \sigma_R$$

$$\frac{1}{R^2}\frac{dM}{d\theta} - \frac{1}{R}\frac{dN}{d\theta} = \tau_{R\theta} \qquad (2.11)$$

Integration of Eq (2.11) gives the expression of $N$ and $M$ versus $p, q, r$.

$$N = pR + \frac{1}{3}(2t - q)R\cos 2\theta$$

$$M = \frac{1}{6}R^2(t - 2q)\cos 2\theta \qquad (2.12)$$

The stiffness moduli of a support depend on the dimension of the section. For a circular tunnel, the normal stiffness is inversely proportional to the tunnel radius and the bending stiffness is inversely proportional to the cube of the radius.

The stiffness moduli of the supports cannot be considered as soft or rigid whatever the ground. The same support can be rigid in soils or weak rocks and soft in hard rocks.

## 2.4 Timber Supports

Timber supports have been used for a long time especially in mines. The simple frames consisted of two posts and a cap. The excavation of large tunnels in parted section needed more complex wooden structures (Fig. 2.4).

In unstable ground a complete or partial sheeting by wooden boards is placed behind the frames.

Timber supports are rarely used today. They are difficult to place and require a highly skilled manpower. They stand poorly in damp atmosphere and rot. They have the advantage to forewarn the failures by cracking.

## 2.5 Steel Ribs

Steel ribs have progressively replaced wooden supports. They are made by the assembling of bent metallic elements to match the excavation profile.

There are different types of profiles: HEB, IPN, TH (Fig. 2.5 and Table 2.1).

**Fig. 2.4** Gotthard Railway
Tunnel. Wooden support of
the upper part

**Fig. 2.5** Cross sections TH
and HEB of steel ribs

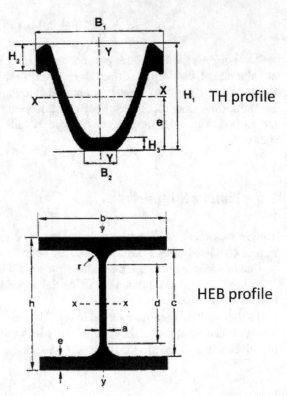

Lattice girders (Fig. 2.6) are often used because they are lighter and have a large moment of inertia. The composite support of lattice girders and sprayed concrete are now common. The sprayed concrete entirely coats the lattice girders (Table 2.2).

Steel ribs are usually made of several elements. Rigid steel ribs are assembled by bolting of plates welded at the extremity of each element or by fish plating. Sliding

**Table 2.1** Characteristics of some common profiles of steel ribs (standard AFNOR)

| HEB profile | Height H (mm) | Width B (mm) | Lineal mass M (kg/m) | Crosssection A (cm²) | Moment of inertia $I_x$ (cm⁴) | Strength Modulus $\frac{I_x}{v_x}$ (cm³) |
|---|---|---|---|---|---|---|
| 120 | 120 | 120 | 26.7 | 34 | 864 | 144 |
| 140 | 140 | 140 | 33.7 | 43 | 1509 | 216 |
| 160 | 160 | 160 | 42.6 | 54.3 | 2492 | 311 |
| 180 | 180 | 180 | 51.2 | 65.3 | 3831 | 416 |
| 200 | 200 | 200 | 61.3 | 78.1 | 5696 | 570 |
| 240 | 240 | 240 | 83.2 | 106 | 11,260 | 938 |

| IPN profile | H (mm) | B (mm) | M (kg/m) | A (cm²) | $I_x$ (cm⁴) | $\frac{I_x}{v_x}$ (cm³) |
|---|---|---|---|---|---|---|
| 100 | 100 | 50 | 8.32 | 10.6 | 171 | 34.2 |
| 120 | 120 | 58 | 11.2 | 14.2 | 328 | 54.7 |
| 140 | 140 | 66 | 14.4 | 18.3 | 573 | 81.9 |
| 160 | 160 | 74 | 17.9 | 22.8 | 935 | 117 |

| TH profile | H (mm) | B (mm) | M (kg/m) | A (cm²) | $I_x$ (cm⁴) | $\frac{I_x}{v_x}$ (cm³) |
|---|---|---|---|---|---|---|
| 21/58 | 108 | 127 | 21 | 27 | 324 | 60 |
| 29/58 | 124 | 150 | 29 | 37 | 598 | 93 |
| 36/58 | 138 | 171 | 36 | 46 | 972 | 137 |
| 44/58 | 145 | 171 | 44.3 | 56.4 | 1175 | 150 |

| Type of steel | Limit of elasticity (N/mm²) | Tensile strength (N/mm²) | Limit strain (%) |
|---|---|---|---|
| TH. 350 | ≥350 | ≥550 | ≥18 |
| TH. 520 | ≥520 | ≥650 | ≥19 |

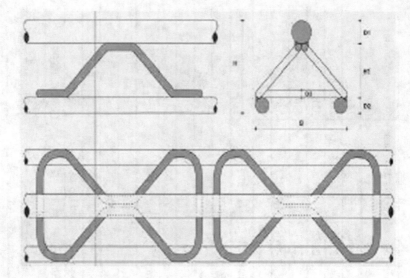

**Fig. 2.6** Cross section of lattice girders

**Table 2.2** Characteristics of lattice girders PANTEX (UNIMETAL)

| Type | | | | | | | |
|---|---|---|---|---|---|---|---|
| $H_1$ (mm) | $d_1$ (mm) | $d_2$ (mm) | H (mm) | B (mm) | M (kg/m) | $I_x$ (cm$^4$) | $\frac{I_x}{v_x}$ (cm$^3$) |
| 70 | 20 | 30 | 120 | 140 | 12.5 | 306 | 51 |
| 70 | 26 | 34 | 130 | 140 | 17.5 | 501 | 71 |
| 110 | 20 | 30 | 160 | 220 | 14.1 | 612 | 75 |
| 110 | 26 | 34 | 170 | 220 | 19.2 | 971 | 105 |
| 130 | 20 | 30 | 180 | 220 | 14.1 | 805 | 87 |
| 130 | 26 | 34 | 190 | 220 | 19.2 | 1264 | 122 |

steel ribs may slide with an overlap of about fourteen centimeters. The sliding is controlled by the tightening of two yokes (Table 2.3).

In cohesionless soils and very fractured rocks, to prevent the flowing of the ground between the ribs, a partial or complete sheeting by wooden boards or metallic plates is placed behind the ribs. Shotcrete reinforced by a wire mesh more and more replaces the sheeting by plates.

**Table 2.3** Characteristics of the assembly of sliding steel sets TH (data UNIMETAL)

| Sections TH | TH 21/58 | TH 29/58 | TH 36/58 | TH 44/58 |
|---|---|---|---|---|
| Tightening torque (Nm) | 350 | 500 | 500 | 500 |
| Sliding load (kN) Laboratory tests on straight elements | 120 | G 405 (*) 180 G 445 (*) 250 | G 405 (*) 250 G 445 (*) 350 | G 445 (*) 350 |

Longitudinal struts between the ribs avoid longitudinal instability.

The efficiency of the support by ribs depends on the quality of the blocking of the ribs against the ground. A discontinuous blocking of the ribs may be achieved by wooden wedges. One or two layers of shotcrete ensure a continuous blocking.

The stiffness characteristics of a support by ribs depend on:

- the cross section of the ribs,
- the shape of the ribs,
- the spacing between two ribs,
- the blocking of the ribs against the ground.

For circular ribs blocked continuously against the ground with a spacing $p$, the normal stiffness and the bending stiffness are given by:

$$K_{sn} = \frac{E_a A}{p R}$$

$$K_{sf} = \frac{E_a I}{p R^3} \tag{2.13}$$

where

$A$    is the area of the cross section,
$E_a$   is the Young's modulus of the steel,
$I$     is the moment of inertia of the cross section.

When the ribs are blocked against the ground in a discontinuous manner by $n$ wedges regularly spaced (Fig. 2.7) Hoek and Brown (1980) have proposed the following expression for the normal stiffness:

**Fig. 2.7** Steel ribs blocked against the ground by spaced blocks

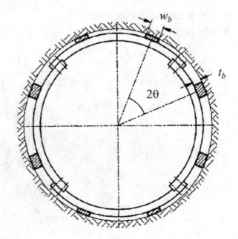

$$\frac{1}{K_{sn}} = \frac{pR}{E_a A} + \frac{pR^3}{E_a I}\left[\frac{\theta(\theta + \sin\theta\cos\theta)}{2\sin^2\theta} - 1\right] + \frac{2p\theta t_b}{E_b w_b^2} \qquad (2.14)$$

where,

| | |
|---|---|
| $2\theta$ | is the angle between two consecutive wedges, |
| $E_b$ | is the Young's modulus of the wedges material, |
| $w_b$ and $t_b$ | are the width and the height of the wedges. |

In the case of a sliding rib, the support confinement curve is given by an elasto-plastic curve, the plastic yield being fixed by the normal load which causes the slide.

## 2.6 Bolting

Rockbolts were first used in mining in the 1890s. They were applied later in civil engineering after the Second World War. They brought about great improvements in the underground works. The manufacturers offer various types of bolts. It is possible to distinguish four types (Fig. 2.8).

(a)    the point anchored bolts,
(b)    the grouted bolts bound on their whole length and non-tensioned,
(c)    the bolts tensioned then grouted,
(d)    the friction bolts such as Split Set and Swellex.

The advantages and the drawbacks of each type of bolts are to be analyzed in each case. The choice of the most adequate type is based on the following factors:

- the nature of the ground,
- the simplicity of handling,

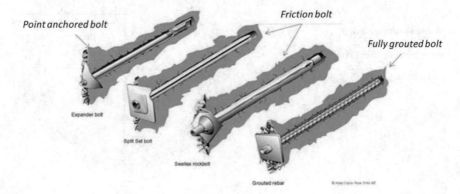

**Fig. 2.8** Various types of bolts (from recommendations AFTES)

- the immediate efficiency of the bolting,
- the quality control,
- the resistance again blasting vibrations,
- the resistance to corrosion.

The main parameters of a bolting pattern are: the bolt type, the material (steel, resin, fiber glass), the rod diameter, the initial tension, the surface plate and the mesh, the spacing of the bolts.

The first use of bolts was to prevent rockfalls in fractured rock masses, but its use has spread to a great variety of grounds with different purposes:

- to carry out a reinforced ground zone around the excavation in fractured rock masses,
- to prevent the bending of layers in stratified rock masses,
- to prevent the buckling failures in schistose rocks,
- to control the convergences,
- to control the extrusion of the face,
- to support the failed zone around the excavation.

Actually, the ground bolting has three major actions: to support failed zones, to prevent local failures, to confine the ground around the excavation (Panet 1991). In the convergence-confinement method, it is the confinement action which is considered.

The maximum load of a bolt is usually given by the rod which must break before the failure of other components (anchoring, surface plate). It may be checked by in situ pull-out tests.

The usual characteristics of common bolts are given in Table 2.4.

The stiffness characteristics of a support by bolting depend on the type of the bolts. For end-anchored bolts, the deformation of the rod is constant between the surface plate and the anchor. Any local displacement of the ground which occurs between the plate and the anchor, such as at the intersection of a discontinuity, results in a uniform change of the stress along the rod. For grouted bolts, a local deformation of the ground modifies locally the stress distribution along the rod. Therefore, the stiffness of an end-anchored bolting is much smaller than the stiffness of a grouted bolting. In ground conditions with large convergences, an end-anchored bolting must be preferred to a grouted bolting. The use of steel bolts with a high failure strain, up to 20%, is recommended. The capacity of bolting to control large convergences may still be improved by putting crushable wooden blocks between the surface plate and the ground.

For end-anchored bolting, Hoek and Brown (1980) have proposed the following expression for the normal stiffness:

$$\frac{1}{K_{sn}} = \frac{e_T e_L}{R} \left( \frac{4L}{\pi d^2 E_b} + Q \right) \tag{2.15}$$

**Table 2.4** Mechanical characteristics of common bolts

| Common bolts | | Type of steel | Young's modulus E (GPa) | Elastic Limit (MPa) | Elastic Limit Load $T_e$ (kN) | FailureLoad Tr (kN) | Failure Strain (%) | Shear Strength |
|---|---|---|---|---|---|---|---|---|
| Friction bolts | Swellex | S275JR/S355MC | 210 | 275–355 | 90–190 | 110–240 | 10–20 | 400 MPa |
| | SplitSet | | 210 | 500 | 30–90 | 105–150 | 15 | 350 MPa |
| Rods for end-anchored bolts and grouted bolts | HA25 | FeE500 | 210 | 500 | 210 (thread)–246 | 230 (thread)–270 | 12 | 180 kN |
| | HA32 | FeE500 | 210 | 500 | 347 (thread)–402 | 382 (thread)–442 | 12 | 293 kN |
| | Self-drivingBolts ($\phi_{ext}$:27–100 mm) ($\phi_{int}$:16–78 mm) | E355–E460 | 210 | 470–590 | 180–2700 | 220–3460 | | 58–88 kN |
| Fiberglass Bolts | Solid or hollow round rods, Y-section,flat rods | – | 40 | 750–1000 | – | – | 3–4 | 100–200 MPa |
| Carbon Bolts | Solid round rods or flat rods Sections of 44 mm to 200 mm$^2$ | – | 130 | 2300 | 100–450 | 100–450 | 1.8 | – |
| Grouted End-anchored Bolts | Φ 20–22 mm | FeE500 | 210 | 500 | 157–190 | 173–266 | | |

where

$e_T$ is the transversal bolt spacing,
$e_L$ is the longitudinal bolt spacing,
$L$ is the length of the bolts,
$d$ is the bolt diameter,
$E_b$ is the Young's modulus of the bolt material,

$Q$ is a parameter taking into account the displacements of the bolt anchor and of the surface plate. It may be determined from the load–displacement curve obtained by a pull-out test on a bolt (Fig. 2.9).

For a steel end-anchored bolt 3 m long with a 22 mm diameter, the value of $Q$ is comprised between 0.03 m /MN and 0.05 m/MN in a fair quality rock.

In this method introduced by Hoek and Brown, it is considered that the action of the bolting is a support pressure increasing with the convergence. It neglects the stresses induced in the ground by the anchor. They were introduced by Egger (1973) but he did not take into account the variation of the bolt load with the convergence. Labiouse (1994) has considered both the variation of the bolt load with the convergence and the ground stresses due to the anchor. In the convergence-confinement method, the mechanical parameters of the bolted ground zone are thus modified.

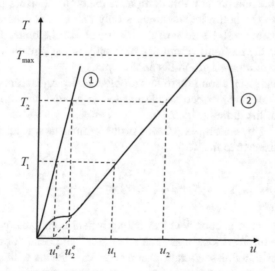

$T$ traction applied on the bolt
$u$ bolt extension

① Load-displacement curve $T = \dfrac{\pi d^2}{4L} E_b u$

② Load-displacement curve of a pull-out test $Q = \dfrac{\left(u_2 - u_2^e\right) - \left(u_1 - u_1^e\right)}{T_2 - T_1}$

**Fig. 2.9** Determination of the $Q$ parameter by an in situ pull-out test on an anchored bolt. After Hoek and Brown (1980)

**Fig. 2.10** Periodic
distribution of bolts at the
wall of the tunnel. After
Greuell et al. (1993)

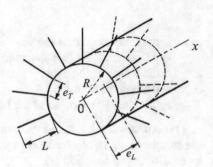

**Fig. 2.10** Periodic distribution of bolts at the wall of the tunnel. After Greuell et al. (1993)

The determination of the Support Confinement Curve of a grouted bolting is more complex. In fact, the bolts must be considered as internal reinforcement elements of the ground that decrease its deformability and increase its strength. The reinforced zone may be considered as a composite material. The mechanical properties of this composite material depend on the properties of the components, the ground and the bolts, and of their distribution. The properties of this equivalent composite medium may be analyzed by homogenization techniques. The usual regular distribution of the bolts in the ground allows to refer to the theory of homogenization of periodic media. Because of the bolt pattern, this composite medium is anisotropic.

It is assumed that the bolt reinforcement is only uniaxial. It decreases with the distance from the tunnel wall because of the divergence of the bolts. The bolt reinforcement increases the modulus of deformation only in the direction of the bolts. The cohesion of the reinforced ground is anisotropic.

Greuell (1993) has given a solution of this problem for a circular tunnel with radial bolts (1993). The bolts are perfectly adherent to the ground and are periodically spaced at the wall of the tunnel (Fig. 2.10).

The reinforcement by bolting is present in the ground in a volume proportion depending on the distance to the wall:

$$\text{for } R \leq r < R + L, \quad a(r) = \frac{\pi d^2}{4} \frac{1}{e_L e_T} \frac{R}{r} \tag{2.16}$$

The reinforced ground is considered as an equivalent homogeneous continuous medium. If the unreinforced rock mass has an elastoplastic behavior characterized by a Young modulus E, a Poisson's ratio $v$ and a Tresca criterion with a cohesion $C$, the reinforced rock mass is represented as an equivalent anisotropic medium with the following elastic characteristics are:

$$E_r = E(1 + K(r))$$

$$E_\theta = \frac{E(1 + K(r))}{(1 - v^2)K(r) + 1}$$

$$v_{\theta r} = v$$

$$v_{\theta x} = v \frac{(1 + v)K(r) + 1}{(1 - v^2)K(r) + 1}$$

$$G = \frac{E}{2(1 + v)} \tag{2.17}$$

where

$$K(r) = \frac{\pi d^2}{4} \frac{E_b}{e_L e_T} \frac{1}{E} \frac{R}{r} \tag{2.18}$$

and $E_b$ is the Young's modulus of the bolt material.

In the elastic domain, the main contribution of the bolting is to increase the modulus of deformability in the bolt direction. The term, $\frac{\pi d^2}{4} \frac{E_b}{e_L e_T}$ characterizes this effect.

For a bolting density of $1/m^2$ with 20 mm diameter steel bolts, it is equal to 65 MPa. In these conditions, if the rock mass has a modulus of deformation larger than 700 MPa, this effect is lower than 10% and may be neglected in the elastic domain.

Considering an elastoplastic medium with a Tresca yield criterion, Greuell (1993) has proposed an evaluation of the cohesion of the reinforced ground by applying the homogenization method and using a limit analysis approach. The cohesion $C_{\text{hom}}(\alpha)$ is anisotropic and the Tresca criterion of the equivalent homogenized medium is:

$$\sigma_1 - \sigma_3 = 2C_{\text{hom}}(\alpha) \tag{2.19}$$

where $\alpha$ is the angle between the bolt direction and the minor principal stress (positive compression stress).

The cohesion $C_{\text{hom}}(\alpha)$ is given by the following expressions:

$$\text{For } 0 \leq \alpha \leq \alpha_T, \quad C_{\text{hom}}(\alpha) = \frac{1}{2}t(r) \cos 2\alpha + C \sqrt{1 - \left(\frac{t(r)}{2C} \sin 2\alpha\right)^2}$$

$$\text{For } \alpha_T \leq \alpha \leq \frac{\pi}{4}, \quad C_{\text{hom}}(\alpha) = \frac{C}{\sin 2\alpha}$$

$$\text{For } \frac{\pi}{4} \leq \alpha \leq \frac{\pi}{2}, \quad C_{\text{hom}}(\alpha) = C \tag{2.20}$$

where

$$\tan 2\alpha_T = \frac{2C}{t(r)}, \quad t(r) = \frac{T_b}{e_L e_T} \frac{R}{r} \tag{2.21}$$

$T_b$ is the tensile strength of the bolt.

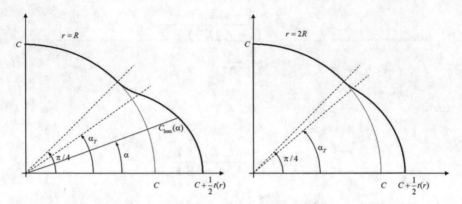

**Fig. 2.11**  Polar distribution of the anisotropic cohesion of the homogenized medium. After Greuell (1993)

The increase of the cohesion decreases with the distance from the wall of the tunnel. It is maximal for $\alpha = 0$ (Fig. 2.11).

$$C_{\text{hom}}(0) = \frac{1}{2}t(r) + C \tag{2.22}$$

The product of the tensile strength by the bolting density characterizes the improvement of the cohesion. For $T = 150$ kN, a bolting density of $1/\text{m}^2$ brings about a maximal increase of cohesion at the wall of 75 kPa.

These results are in agreement with the experimental data obtained by Egger (1973) and Wüllschläger and Natau (1987).

Bernaud et al. (1995a, b) carried out a similar approach for an elastoplastic rockmass with a Drucker-Prager criterion.

## 2.7   Shotcrete

The use of shotcrete as a support of a tunnel was mainly developed with the New Austrian Method (NATM). The NATM was introduced by an Austrian engineer (Rabcewicz 1964, 1965). The NATM is based on theoretical concepts which are not well defined and have changed with time and with the nature of the ground. Nevertheless, the use of a 50–200 mm layer of shotcrete to provide an immediate ground reinforcement close to the face is widely spread all over the world. A fair combination of shotcrete, wire mesh and bolting is a very common flexible and adaptable support.

The shotcrete has the composition of a pumpable concrete. It may be dry-prayed, with water added to the dry components at the nozzle, or wet-sprayed, with ready

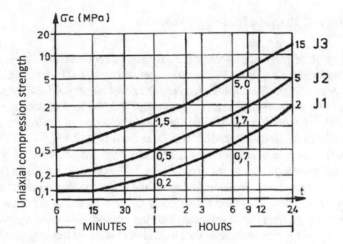

**Fig. 2.12** Austrian specifications—uniaxial compressive strength of the shotcrete at early age (Austrian Concrete Society 1990)

mixed concrete pumped to the nozzle. Dry spraying was initially preferred, but nowadays wet spraying is more used. Its advantages are:

- better quality control of the water/cement ratio,
- less rebound during spraying,
- healthier atmosphere on the working site,
- more economical technique.

The choice of the admixtures (plasticizer, accelerator …) and of the additives allows to adjust the desired time setting, a sufficient resistance during the hardening and the long term characteristics. Standards have been published in various countries (Fig. 2.12).

Incorporation of steel or plastic fibers improves the shotcrete characteristics, especially the resistance to large deformations.

During the setting of the shotcrete, the characteristics of deformability and strength increase with time. In the convergence-confinement method, the shotcrete layer is considered as a structural ring. Its stiffness increases with time. For the sake of simplicity, an average stiffness at early age is often introduced. A value of about 8000 MPa may be suggested. For the final support, the long-term characteristics of the shotcrete are taken into account.

The confinement of the ground may induce the failure of the young shotcrete in grounds prone to large convergences. Several successive layers of shotcrete may be applied to face these failures. These situations may be analyzed by the convergence-confinement method.

## 2.8   Rings of Prefabricated Segments

A support made of rings of prefabricated cast iron or reinforced concrete segments is commonly in tunnels with a circular cross section excavated by a tunneling boring machine. Depending on the tunnel diameter, the rings comprise 5–10 segments.

A support by prefabricated segments may also be used to construct a vault. This is the case of the Jacobson vaults used for the construction of the Auber station of RER A and Gare du Nord-Est station of Eole in Paris (Fig. 2.13).

For circular tunnels excavated by a TBM, the rings may be erected behind the shield tail and expanded against the ground (Fig. 2.14) or under the shield tail. In this case, the annular gap between the outer face of the ring and the ground must be filled by a mortar (Fig. 2.15). This mortar is usually not considered in the analysis of the interaction between the ring of segments and the ground. The grouting pressure of this mortar creates a prestressing of the ring. This loading of the ring may be larger than the ground pressure.

The rings are assembled by transversal joints and the segments of a ring by longitudinal joints. The transversal joints and the longitudinal joints are different because of the different loads that they support.

Rings comprise usually rectangular segments with a key segment and two trapezoidal counter segments adjacent to the key segment. Different types of rings with trapezoidal segments may be designed. The rings with the universal segments allow to comply with the tunnel alignment.

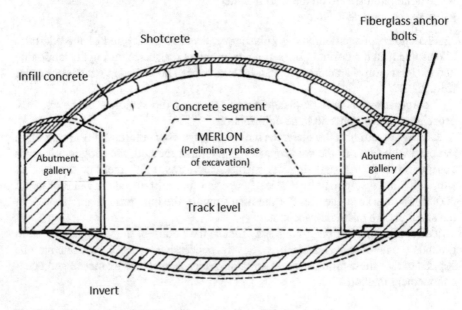

**Fig. 2.13**  Cross section of the Nord-Est station of the Eole line in Paris

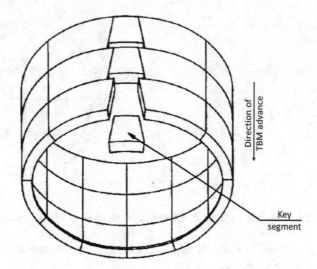

**Fig. 2.14** Rings of segments expanded against the ground

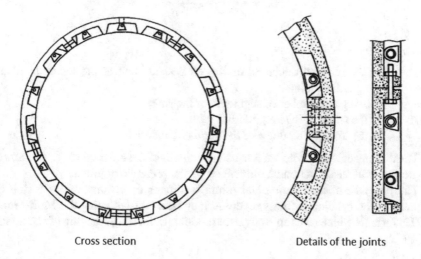

Cross section                                    Details of the joints

**Fig. 2.15** Rings of bolted segments

To evaluate the stiffness of a ring of segments, an equivalent cylindrical shell is considered. The behavior of the longitudinal joints is to be taken into account.

There are three types of longitudinal joints:

- cylindrical contact joints, convex-convex or concave-convex,
- combined geometry contact joints which are usually bolted,
- plane contact joints with bolts or connectors.

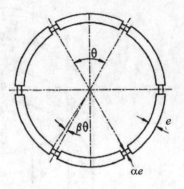

**Fig. 2.16** Modeling of a ring of $n$ segments

They can be modelled as a zone of reduced thickness of the cylindrical shell (Fig. 2.16).

The normal stiffness is given by Eq. 2.3, $e$ being the thickness of the segments and $E_s$ an equivalent modulus given by;

$$E_s = \frac{\alpha}{\alpha(1 - \beta) + \beta} E_m \tag{2.23}$$

where

$\alpha e$      is the reduced thickness of the equivalent shell at the location of the longitudinal joints,

$\theta = \frac{2\pi}{n}$    $n$ being the number of segments of the ring,

$\beta\theta R$      is the width of the longitudinal joint,

$E_m$      is the Young's modulus of the segment material.

The dimensionless coefficient $\beta$ is usually small, of the order of $10^{-3}$. Therefore, the normal stiffness is not much influenced by the presence of joints.

The presence of the longitudinal joints facilitates the ovalization of a ring of segments. The bending stiffness modulus may be evaluated using the Muir-Wood (1972) formula which gives an equivalent modulus of inertia of the ring of segments:

$$I = I_j + \left(\frac{4}{n}\right)^2 \frac{e^3}{12} \tag{2.24}$$

where

$I_j$    is the modulus of inertia of the joints,

$n$    is the number of joints.

This formula is valid if $n > 4$.

Assuming that the moment of inertia of the joints is:

$$I_j = \frac{\alpha^3 e^3}{12} \tag{2.25}$$

then,

$$I = \left[\alpha^3 + \left(\frac{4}{n}\right)^2\right]\frac{e^3}{12} \tag{2.26}$$

For example, if $\alpha = 0.5$ and $n = 6$: $I = 0.57\frac{e^3}{12}$.

The presence of joints reduces the bending stiffness and the bending moments.

Muir Wood's formula assumes that the joints of two consecutive rings are aligned. If they are not, the bending stiffness of the support is larger than that derived from Muir Wood's formula.

## 2.9  Yielding Elements

During the construction of the first Tauern Tunnel in Austria, strong convergences occurred and caused some failures of the shotcrete shell. To avoid these failures, L. Rabcewicz, the father of NATM (1971–1974), had horizontal slots made in the shotcrete shells in order to reduce the rigidity. To control the closure of the shells in similar conditions, various devices to be inserted in the slots were later designed. Prefabricated highly deformable concrete blocks are presently the most advanced solution. They consist of high strength concrete with porous aggregates reinforced by steel fibers, stirrups and plates (Fig. 2.17). The closing of the aggregates pores induced by the applied compression load allows a plastic deformation up to 40% (Fig. 2.18).

## 2.10  Composite Supports

In many cases several types of support are associated:

- bolts and shotcrete,
- ribs and shotcrete,
- ribs, bolts and shotcrete.

In the analysis of the ground-support interaction, it is often considered that all the components of the support are installed simultaneously. The stiffness of the composite support is equal to the sum of the stiffness of the components. But if the placement of two supports is shifted, the SCC is obtained from the characteristic curves of each component as shown in Fig. 2.19.

It may be noticed that the stiffness of the various types of support may be very different. Table 2.5 gives the normal stiffness modulus of five types of support for a tunnel with circular section of 10 m diameter.

**Fig. 2.17**  Tunnel Lyon-Turin—Saint-Martin-La-Porte gallery—support by sliding ribs, schotcrete shell with yielding blocks hiDCon (Courtesy of Eiffage)

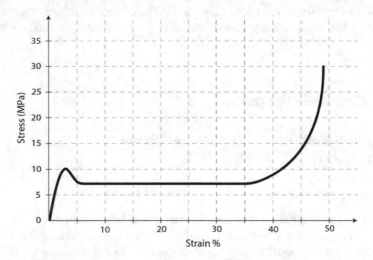

**Fig. 2.18**  Yielding element hiDCon. Compression test (Courtesy of Solexperts)

The stiffness of some types of support appears to be very low compared to others components and may be neglected in the convergence-confinement method. However, they may be necessary to insure the stability of the roof and the walls and a continuous behavior of the ground.

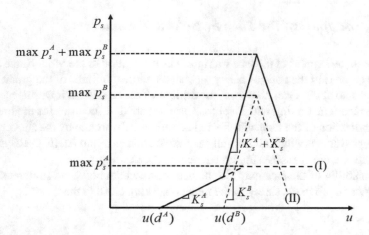

**Fig. 2.19** Support confinement curve of a composite support

**Table 2.5** Normal stiffness of five types of support for a tunnel with circular section of 10 m diameter

| Type of support | a | b | c | d | e |
|---|---|---|---|---|---|
| $K_{sn}$ (MPa) | 2500 | 1800 | 210 | 190 | 22.5 |
| $K_{sf}$ (MPa) | 1.35 | 0.5 | $4 \times 10^{-3}$ | $2.5 \times 10^{-3}$ | 0 |

(a) concrete cast concrete ring 40 cm thick; (b) ring of 6 reinforced concrete segments 30 cm thick; (c) shotcrete shell 10 cm thick (early age); (d) circular ribs HEB140 spaced 1 m apart; (e) anchor bolts 4 m long with a 18 mm rod diameter and a bolting density of $1/m^2$

## 2.11 Two-Phase Supports

In squeezing grounds with large convergences, some severe yielding of the support may occur behind the face. A complementary support, called the second phase support, is necessary to face the damage of the first phase support. In some cases, when the tunnel cross section is no more met, it is necessary to remove the damaged support.

This two-phase support can be introduced in the convergence-confinement method.

## 2.12 Support of the Face

Face instabilities may occur in difficult ground conditions. To face these instabilities or limit the extrusion of the face, a face support is necessary. In the convergence-confinement method the face support is to be considered for the determination of the longitudinal profile of displacement (LPD) and of the deconfinement ratio.

### 2.12.1   Support of the Face by Bolts and Shotcrete

In the past, the support of the face was rare. The excavation of the Mont Blanc Tunnel may be quoted. In the zone of heavy rockbursts, violent failures of the granitic rock occurred also at the face. To permit the drilling of the next shift, it was necessary to bolt the face. The bolting with steel rods brought about difficulties for the mucking. The stabilization of the face with fiber glass bolts easily broken by the shovels was a great improvement which allows full-face excavation as in the ADECO-RS method. Face bolts are the cheapest support element at the heading.

The stability of the face may also be obtained by both bolts and shotcrete. These techniques improve the safety of the miners working close to the face.

### 2.12.2   Ground Treatments Ahead of the Face

In order to cross very difficult zones such as geological faults, grouting of the ground is commonly used. In water-bearing zones, grouting is combined with drainage by boreholes.

The temporary ground reinforcement may be insured by ground freezing. The ground freezing is obtained by the circulation of brine or liquid nitrogen in closed pipes. The concept of ground freezing was first introduced in mining works in France. The temperature of the ground freezing by the brine (calcium chloride solution) is between $-15$ and $-25$ °C. The liquid nitrogen is produced industrially at a temperature of $-196$ °C under the atmospheric pressure. The ground freezing may cause difficulties due to the swelling of the ground during the freezing or to the thaw of the frozen ground.

### 2.12.3   Confinement of the Tunnel Face

The confinement of the face in a compressed air chamber dates back from the nineteenth century, for the crossing of water-bearing zones or for subaqueous tunneling.

Nowadays, the tunnel boring machines may be equipped to confine the face by air pressure, slurry, or by the excavated earth maintained in the cutterhead chamber.

The confinement pressure applied to the face is to be introduced in the convergence-confinement method to determine the deconfinement ratio.

## 2.13  Presupport

As shown above, the support is post often installed at an unsupported span from the face. However, to stabilize the face during construction and to reduce the preconvergence, a presupport may be necessary.

Just as a distance of influence $d_1$ is defined behind the face, a distance of influence ahead of the face $d_2$ may be defined. It may be determined from the extrusion measurements. Most of the time this distance is less than the tunnel span. The useful length of the presupport may be evaluated to 1.5 times the tunnel span. But for practical reasons, the presupport length is often much larger.

The most traditionnal method of presupport is forepoling (Fig. 2.20). Spiles drilled ahead of the face provide a canopy to enable the heading of the tunnel to be advanced safely. The spiles are propped against the ribs set close to the face. Forepoling significantly reduces the excavation rate and the support ahead of the face is less than 2 m.

In the so-called "*umbrella vaults*", horizontal columns are drilled from the face (Fig. 2.21). The dips of the boreholes are from 8° to 10° The cover between successive vaults is comprised between 1 and 3 m. Heavy steel ribs spaced from 1 to 3 m and placed close to the heading support the columns.

The horizontal columns often consist of grouted tubes spaced between 20 and 50 cm apart. The horizontal columns are not jointed and for cohesionless ground, raveling may occur between the columns. The vault effect is not very effective and the columns may be considered as horizontal beams supported by the ground ahead of the face and by the steel sets behind the face.

Columns of jet-grouting are now commonly used to make an umbrella vault. The columns diameter depends on the nature of the soil and varies between 20 and 60 cm. They may be considered as contiguous and they act as a vault.

Perforex has developed an original presupport method, known as mechanical precutting. A slot is cut ahead of the face and filled with a fiber reinforced sprayed concrete (Fig. 2.22). The slot is 14–20 cm wide and 3 m deep. The stability of the slot after cutting is the main limit of the use of mechanical precutting. The prevaults are slightly conical with a cover variable with the nature of the ground.

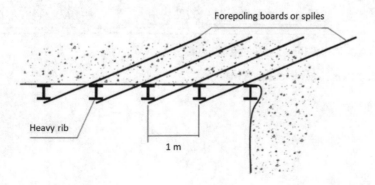

**Fig. 2.20**  Presupport by forepoling

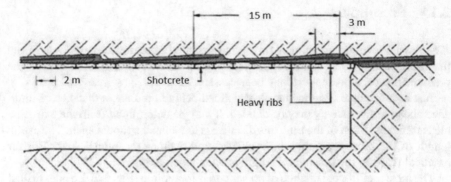

**Fig. 2.21**  Presupport by grouted horizontal columns

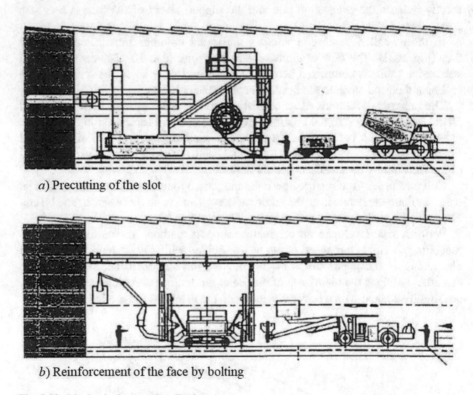

*a*) Precutting of the slot

*b*) Reinforcement of the face by bolting

**Fig. 2.22**  Mechanical precutting (Perforex)

## 2.14 Lining

In most tunnels the temporary support is generally completed by the final permanent lining.

In old tunnels a masonry lining was used. Nowadays a cast in place concrete lining comprising only a vault or a vault and a counter vault is most commonly used. In tunnels driven by a TBM, the rings of segments may constitute the final lining.

The cast in place concrete linings are usually not reinforced; They may be locally reinforced at the junction of the vault and the invert.

For a long time, the final lining was designed to bear the actions supported by the preliminary support plus the actions brought about by the differed ground behavior. Due the durability of the modern supports, it is now assumed that the preliminary support may also be active in long term.

In the convergence-confinement method, the design of the final lining is done considering the ground reaction curve assuming the long term mechanical parameters of the ground.

## References

Bernaud D, De Buhan P, Maghous S (1995a) Numerical simulation of the convergence of a bolt-supported tunnel through a homogenization method. Int J Numer Anal Meth Geomech 19(4):267–288

Bernaud D, De Buhan P, Maghous S (1995b) Calcul numérique des tunnels boulonnés par une méthode d'homogénéisation. Rev Fr Géotech 73:53–65

Bieniawski ZT (1989) Engineering rock mass classifications: a complete manual for engineers and geologists in mining, civil and petroleum engineering. Wiley, Hoboken, NJ, USA

Bieniawski ZT (1993) Classification of rock masses for engineering: The RMR system and future trends. In: Comprehensive rock engineering, Pergamon Press, London, vol 3, 22 pp 553–573

Egger P (1973) Einfluss des Post Failure Verhaltens von Fels auf den Tunnellaubau. Thèse. Veröff Inst. Bodenmechanik, Karlsruhe

Greuell E (1993) Etude du soutènement par boulons passifs dans les sols et les roches tendres par une méthode d'homogénéisation. Thèse de l'Ecole Polytechnique, Paris

Greuell E, De Buhan P, Panet M, Salençon J (1993) Comportement des tunnels renforcés par boulons passifs. In: Proceedings of XIII international conference on soil mechanics and foundation engineering, New-Dehli, India

Hoek E, Brown ET (1980) Underground excavations in rock. Institution of Mining and Metallurgy, London

Labiouse V (1994) Etude par "convergence-confinement" du boulonnage à ancrage ponctuel comme soutènement de tunnels profonds creusés dans la roche. Rev Fr Géotech 65:17–28

Muir-Wood AM (1972) The circular tunnel in elastic ground. In: Proceedings of international symposium underground openings, Lucerne, pp 356–369

Panet M (1991) Ground reinforcement by bolts in tunnelling. Convegno "Il consolidamento suolo a delle rocce nelle realizzazioni in sotterraneo", vol 1, pp 31–41

Rabcewicz L (1964) The new Austrian tunnelling method, part one, Water power, Nov 1964, 453–457, Part two, Water power, Dec 1964, 511–515

Rabcewicz L (1965) The new austrian tunnelling method, part three. Water Power, Jan 1965, pp 19–24

Wüllschläger D, Natau O (1987) The bolted rock mass as an anisotropic continuum. Material behaviour and design suggestions for rock cavities. In: Proceedings of congress of international society for rock mechanics, vol 1 pp 1321–1324

## Additional Reading

AFTES (1981) Recommandations du groupe de travail n°7. Soutènement et Revêtement. Tunnels et Ouvrages Souterrains. Numéro spécial, avril 1981

AFTES (1993) Recommandations relatives au choix d'un type de soutènement en galerie. Tunnels Et Ouvrages Souterrains 117:60–71

Barton N (2008) Rock mass classification RMR and Q—Setting records. Tunnels and tunneling international, (Feb), pp 26–29

CETU (1998) Dossier pilote des tunnels—Section 4—Méthodes de creusement et de soutènement, Bron

Deere DG, Peck RB, Monsees JE, Schmidt B (1969) Design of lines and support systems. National technical information service, US Department of Commerce

Lauffer H (1958) Gebrigsklassifizierung Für Den Stollenbau. Geologie Bauwesen 74:46–51

Lunardi P (1993) Nuovi criteri di progetto a construzione per una corretta planificazione della opere in sotterraneo. Samater. Convegno su "la realizzazione delli grandi opere in soterraneo", Verona

Lunardi P, Bindi R, Focaracci A (1989) Nouvelles orientations pour le projet de la construction des tunnels dans les terrains meubles. Etudes et expériences sur le préconfinement de la cavité et la préconsolidation du noyau au front. In: Tunnels et micro-tunnels en terrain meuble. Du chantier à la théorie. Presses de l'ENPC, Paris, pp 577–590

Lyons AC (1977) Some developments in segmental tunnel linings designed in the United Kingdom. Underground Space 1(3):173–183

Panek LA (1982) The combined effect of friction and suspension in bolting bedded mine roofs. US Bureau of Mines. Report of Investigation 6139

Pelizza S, Peila D (1993) Soil and rock reinforcement in tunnelling. Tunn Undergr Space Technol 8(3):357–372

Stillborg B (1986) Professional users handbook for rock bolting. In: Series on rock and soil mechanics. Transtech Publications, vol 15

Subrin D (2002) Etudes théoriques sur la stabilité et le comportement des tunnels renforcés par boulonnage. Thèse de l'Ecole Nationale des Travaux Publics

Wong H, Subrin D, Dias D (2000) Extrusion movements of a tunnel head reinforced by finite length bolts—a closed-form solution using homogenization approach. Int J Numer Anal Meth Geomech 24:533–565

# Chapter 3
# The Convergence-Confinement Method for a Tunnel Driven in an Elastic Medium

For a linear-elastic behavior of the medium under axisymmetric conditions, the convergence-confinement method can be easily developed. It is assumed that the rock mass is homogeneous and isotropic and that the initial stress state is also isotropic. In addition, the tunnel of circular cross section must be deep enough so that the variation of the initial stresses can be neglected in the zone concerned by the tunnel excavation, that is to say that the tunnel radius is small compared to the tunnel depth (Fig. 3.1).

Before studying the interaction between the rock mass and the support, the convergence and the displacements field caused by the excavation of an unsupported tunnel are first analyzed.

## 3.1 Displacements Field and Convergence of an Unsupported Tunnel

The *Longitudinal Displacement Profile* (LDP) which represents the distribution of the radial displacement $u_R$ at the wall of a tunnel with circular cross section of radius $R$ *versus* the distance $x$ to the tunnel face can be obtained by numerical modelling.

When the considered section is ahead of the face ($x < 0$ in the rock mass), and $x < -2R$, $u_R$ is practically zero. When the distance of the considered section to the tunnel face is such that $x > 4R$, $u_R$ does not vary anymore and is given by the classical Lamé solution:

$$u_{R,\infty} = \frac{\sigma_0 R}{2G} \tag{3.1}$$

where $G$ is the elastic shear modulus of the medium and $\sigma_0$ is the initial isotropic in situ stress.

For a distance $x$ such that $-2R < x < 4R$, we can write:

**Fig. 3.1** Deep tunnel with circular cross section

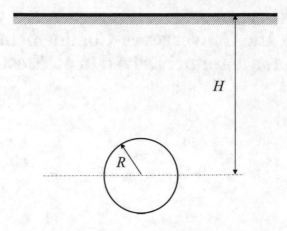

$$u_R(x) = a(x)u_{R,\infty} \tag{3.2}$$

where $a(x)$ is a dimensionless shape function. Explicit expressions for $a(x)$ can be obtained from fitting of finite elements numerical computations. For $x > 0$, the following function is commonly used:

$$a(x) = a_0 + (1 - a_0)\left(1 - \left[\frac{mR}{mR + x}\right]^2\right) \tag{3.3}$$

where $a_0$ is the value of the function $a(x)$ at the tunnel face ($x = 0$). This value depends upon the Poisson's ratio of the medium. However, in practice, it is assumed that the values of the parameters $a_0$ and $m$ are independent of the Poisson's ratio and the current values used are: $a_0 = 0.25$ and $m = 0.75$ or or $a_0 = 0.27$ and $m = 0.84$. These values give the best approximation of the function $a(x)$ for $x > 0.2R$.

More generally, the shape function $b(x)$ as defined by Eq. (3.4) will be used in the following:

$$b(x) = \frac{u_R(x) - u_R(0)}{u_{R,\infty} - u_R(0)} \tag{3.4}$$

This function takes values between 0 for $x = 0$ and 1 for $x \to +\infty$(Fig. 3.2).

Considering that the initial measurement is made at the working face, the convergence is given by:

$$C(x) = 2(u_R(x) - u_R(0)) = (a(x) - a_0)\frac{\sigma_0 R}{G} \tag{3.5}$$

**Fig. 3.2** Shape function $b(x)$

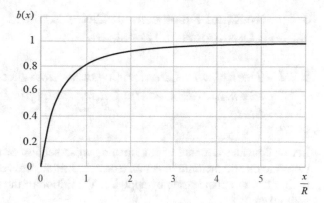

The relative convergence is defined as the dimensionless ratio $c(x) = \frac{C(x)}{2R}$ which can also be written as:

$$c(x) = b(x)c_\infty$$
$$\text{with } c_\infty = (1 - a_0)\frac{\sigma_0}{2G} \tag{3.6}$$

For $x = 0$, the slope of the tangent to the curve $c(x)$ is $\frac{2c_\infty}{mR}$. This tangent thus passes through the coordinate point $\left(\frac{mR}{2}, c_\infty\right)$, which leads to a simple geometric construction to determine the asymptotic value $c_\infty$ (Fig. 3.3).

**Fig. 3.3** Convergence curve

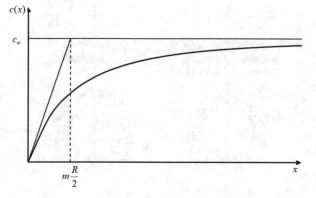

## 3.2  Stress and Displacement Fields Around an Unsupported Tunnel

### 3.2.1  Tunnel with Circular Cross Section Excavated in an Elastic Medium Under Isotropic Initial Stress State

Let us consider a tunnel with a circular cross section of radius $R$ excavated in an isotropic elastic medium under an isotropic initial stress $\sigma_0$. The tunnel excavation is simulated in plane strain by applying a variation of the radial stress at the tunnel wall given by:

$$\Delta\sigma_R = -\lambda(x)\sigma_0 \tag{3.7}$$

where $\lambda(x)$ is the so-called *deconfinement ratio* or *stress release ratio* which depends upon the distance $x$ of the considered section to the tunnel face and represents the progressive relaxation of the pressure applied inside the excavation boundary until the normal stress is effectively zero (Fig. 3.4).

In polar coordinates, the equilibrium equation in plane strain is written as:

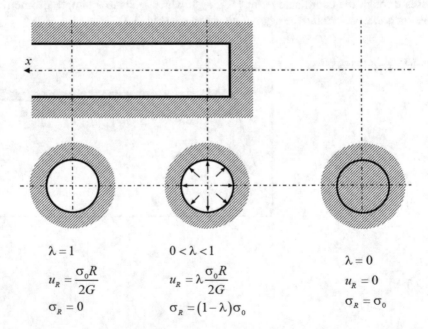

$$\lambda = 1$$
$$u_R = \frac{\sigma_0 R}{2G}$$
$$\sigma_R = 0$$

$$0 < \lambda < 1$$
$$u_R = \lambda\frac{\sigma_0 R}{2G}$$
$$\sigma_R = (1-\lambda)\sigma_0$$

$$\lambda = 0$$
$$u_R = 0$$
$$\sigma_R = \sigma_0$$

**Fig. 3.4** Variation of the *deconfinement ratio* $\lambda$ as a function of the distance to the tunnel face

$$\frac{d\sigma_r}{dr} + \frac{\sigma_r - \sigma_\theta}{r} = 0 \tag{3.8}$$

The stress–strain relationships are given by Hooke's law:

$$\sigma_r = \frac{2G}{1-\nu}(\varepsilon_r + \nu\varepsilon_\theta) = \frac{2G}{1-\nu}\left(\frac{du}{dr} + \nu\frac{u}{r}\right)$$
$$\sigma_\theta = \frac{2G}{1-\nu}(\nu\varepsilon_r + \varepsilon_\theta) = \frac{2G}{1-\nu}\left(\nu\frac{du}{dr} + \frac{u}{r}\right) \tag{3.9}$$

where $G$ is the elastic shear modulus and Poisson's ratio of the medium and $\nu$ is the Poisson's ratio.

By using the boundary conditions of the problem, $\sigma_R = (1 - \lambda)\sigma_0$ and $\lim_{r\to\infty} \sigma_r = \sigma_0$ the following expressions for the displacement, strains and stresses are obtained (Fig. 3.5):

$$u_r = \lambda\frac{\sigma_0}{2G}\frac{R^2}{r}$$

$$\varepsilon_r = -\lambda\frac{\sigma_0}{2G}\frac{R^2}{r^2}$$

$$\varepsilon_\theta = \lambda\frac{\sigma_0}{2G}\frac{R^2}{r^2}$$

$$\sigma_r = (1 - \lambda\frac{R^2}{r^2})\sigma_0 \tag{3.10}$$

$$\sigma_\theta = (1 + \lambda\frac{R^2}{r^2})\sigma_0$$

$$\sigma_x = \sigma_0$$

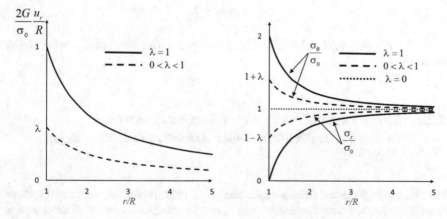

**Fig. 3.5** Isotropic elastic medium, $(K_0 = 1)$—radial displacement and distribution of the principal stresses around the tunnel for various values of the deconfinement ratio $\lambda$.

**Fig. 3.6** Isotropic elastic
medium, $(K_0 = 1)$—stress
and strain paths at the tunnel
wall

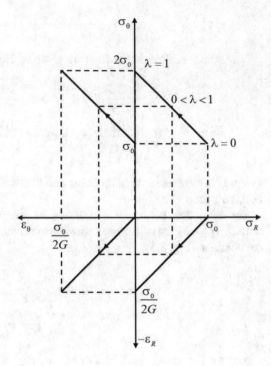

Note that the mean stress and the axial stress are unchanged and equal to $\sigma_0$ and that the deformation occurs at constant volume.

At the tunnel wall, for $r = R$:

$$u_R = \lambda \frac{\sigma_0}{2G} R$$
$$\sigma_R = (1 - \lambda)\sigma_0 \qquad (3.11)$$
$$\sigma_\theta = (1 + \lambda)\sigma_0$$

In Fig. 3.6 the stress and strain paths at the tunnel wall are plotted when $\lambda$ varies from 0 to 1.

### 3.2.2 Tunnel with Circular Cross Section Excavated in an Elastic Medium Under Anisotropic Initial Stress State

For an anisotropic initial stress state, an analytical solution for the stress and displacement fields around a circular section tunnel can be obtained by means of the complex variable method commonly used to solve two-dimensional elasticity problems. The reader can refer to Appendix 1 where the principles of this method are recalled.

It is assumed that the axis of the tunnel corresponds to a principal direction of the tensor of the initial stresses. The initial state of the stresses is homogeneous and anisotropic with a lateral earth pressure coefficient $K_0$. In the plane of a section of the tunnel, the initial stresses are (Fig. 3.7):

$$\sigma = \begin{pmatrix} \sigma_0 & 0 \\ 0 & K_0\sigma_0 \end{pmatrix} \tag{3.12}$$

In cylindrical coordinates, the initial stresses are:

$$\sigma_r^0 = \frac{1}{2}((1 + K_0) - (1 - K_0)\cos 2\theta)\sigma_0$$
$$\sigma_\theta^0 = \frac{1}{2}((1 + K_0) + (1 - K_0)\cos 2\theta)\sigma_0 \tag{3.13}$$
$$\tau_{r\theta}^0 = \frac{1}{2}((1 - K_0)\sin 2\theta)\sigma_0$$

The excavation is simulated by applying a variation of the radial stress $\Delta\sigma_R$ and a variation of the shear stress $\Delta\tau_{R\theta}$ at the tunnel wall such as:

$$\Delta\sigma_R = -\frac{1}{2}\lambda((1 + K_0) - (1 - K_0)\cos 2\theta)\sigma_0$$
$$\Delta\tau_{r\theta} = -\frac{1}{2}\lambda((1 - K_0)\sin 2\theta)\sigma_0 \tag{3.14}$$

**Fig. 3.7** Tunnel with circular cross section under anisotropic initial stress

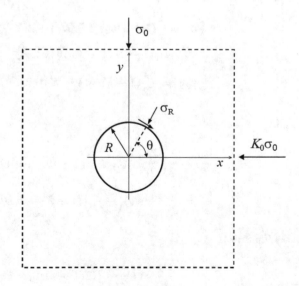

where $\lambda$ is the deconfinement ratio which increases from 0 to 1 as the heading face progresses. Note that it is assumed here that the deconfinement ratio is uniform around the tunnel.

The reader will find in Appendix 1 the mathematical derivations for obtaining the stress and displacement fields as follows:

$$\sigma_r = \frac{1}{2}(1 - \lambda)\sigma_0\left((1 + K_0)\frac{R^2}{r^2} + (1 - K_0)\left(-\frac{4R^2}{r^2} + \frac{3R^4}{r^4}\right)\cos 2\theta\right)$$

$$\sigma_\theta = \frac{1}{2}(1 + \lambda)\sigma_0\left((1 + K_0)\frac{R^2}{r^2} + 3(1 - K_0)\frac{R^4}{r^4}\cos 2\theta\right)$$

$$\sigma_x = \sigma_0\left(1 - 2\lambda(1 - K_0)v\frac{R^2}{r^2}\cos 2\theta\right)$$

$$\tau_{r\theta} = \frac{1}{2}(1 + \lambda)\sigma_0(1 - K_0)\left(\frac{2R^2}{r^2} - \frac{3R^4}{r^4}\right)\sin 2\theta$$

(3.15)

$$u_r = \frac{1}{2}\left(\frac{\lambda\sigma_0}{2G}\right)\left((1 + K_0)\frac{R^2}{r} + (1 - K_0)\left(\frac{R^4}{r^3} - 4(1 - v)\frac{R^2}{r}\right)\cos 2\theta\right)$$

$$u_\theta = \frac{1}{2}(1 - K_0)\left(\frac{\lambda\sigma_0}{2G}\right)\left(\frac{R^4}{r^3} + 2(1 - 2v)\frac{R^2}{r}\right)\sin 2\theta$$

(3.16)

At the tunnel wall for $r = R$

$$\sigma_R = \frac{1}{2}(1 - \lambda)\sigma_0((1 + K_0) - (1 - K_0)\cos 2\theta)$$

$$\sigma_\theta = \frac{1}{2}(1 + \lambda)\sigma_0((1 + K_0) + 3(1 - K_0)\cos 2\theta)$$

$$\sigma_x = \sigma_0(1 - 2\lambda(1 - K_0)v\cos 2\theta)$$

$$\tau_{R\theta} = -\frac{1}{2}(1 + \lambda)\sigma_0(1 - K_0)\sin 2\theta$$

(3.17)

and

$$\frac{u_R}{R} = \frac{1}{2}\left(\frac{\lambda\sigma_0}{2G}\right)((1 + K_0) - (3 - 4v)(1 - K_0)\cos 2\theta)$$

$$\frac{u_\theta}{R} = \frac{1}{2}(3 - 4v)\left(\frac{\lambda\sigma_0}{2G}\right)(1 - K_0)\sin 2\theta$$

(3.18)

For $\theta = 0$

$$\frac{u_R}{R} = \left(\frac{\lambda\sigma_0}{2G}\right)(2(1 - v)K_0 - (1 - 2v))$$

$$\frac{u_\theta}{R} = 0$$

(3.19)

For $\theta = \frac{\pi}{2}$

$$\frac{u_R}{R} = \left(\frac{\lambda\sigma_0}{2G}\right)(-(1-2\nu)K_0 + (1-2\nu))$$

$$\frac{u_\theta}{R} = 0$$

(3.20)

For $K_0 < 1$, the tunnel wall is converging at the crown ($\theta = \frac{\pi}{2}$), whereas if $K_0 < \frac{1-2\nu}{2(1-\nu)}$, the tunnel wall is diverging at $\theta = 0$.

The ground reaction curves are plotted in Fig. 3.8 for $\theta = 0$ and $\theta = \pi/2$, assuming a Poisson's ratio $\nu = 0.25$.

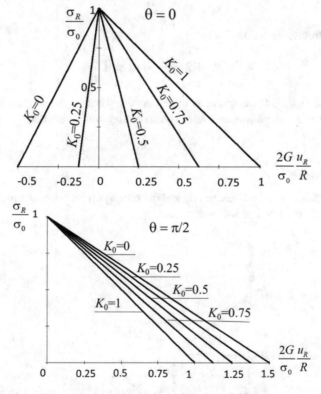

**Fig. 3.8** Ground reaction curves for various values of $K_0$ ($\nu = 0.25$)

## 3.3   Application of the Convergence-Confinement Method

### 3.3.1   Isotropic Initial Stress State

If an elastic support, with normal rigidity $K_{sn}$ is installed at a distance $d$ from the tunnel face ($d$ is called the unsupported span), the application of the convergence-confinement method permits to determine the equilibrium state far from the face (Fig. 3.9). This equilibrium state is simply given by the intersection of the characteristic curve of the support described by the equation:

$$\sigma_R - K_{sn}\frac{u_R - u_d}{R} = 0 \tag{3.21}$$

and of the Ground Reaction Curve:

$$\sigma_R + 2G\frac{u_R}{R} - \sigma_0 = 0 \tag{3.22}$$

In Eq. (3.21), $u_d$ is the displacement that occurred before the support installation. Let $\lambda_d$ denote the deconfinement ratio when the support is installed,

$$\frac{u_d}{R} = \lambda_d\frac{\sigma_0}{2G} \tag{3.23}$$

the support pressure at equilibrium $p_s$ and the corresponding displacement $u_R$ at the tunnel wall are obtained as follows:

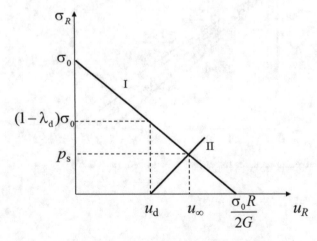

**Fig. 3.9**  Convergence-confinement diagram in the elastic axisymmetric case

$$p_s = \frac{k_{sn}}{1 + k_{sn}}\left(\sigma_0 - 2G\frac{u_d}{R}\right) = \frac{k_{sn}}{1 + k_{sn}}(1 - \lambda_d)\sigma_0$$

$$\frac{u_R}{R} = \frac{1}{1 + k_{sn}}\left(k_{sn}\frac{u_d}{R} + \frac{\sigma_0}{2G}\right)\sigma_0 = \frac{1 + \lambda_d k_{sn}}{1 + k_{sn}}\frac{\sigma_0}{2G} \tag{3.24}$$

where $k_{sn}$ is the relative normal stiffness of the support:

$$k_{sn} = \frac{K_{sn}}{G} \tag{3.25}$$

The main difficulty in applying the convergence-confinement method lies in determining the radial displacement $u_d$ which occurs before the support installation, and therefore in the choice of the corresponding stress release ratio $\lambda_d$. This point will be discussed in detail in Chap. 5.

### 3.3.2 Anisotropic Initial Stress State

In the case of an anisotropic initial stress field, it is no longer possible to define a unique curve characteristic of the response of the ground as we saw in the previous section. It is no longer possible either to define a single characteristic curve for the support. However, the principles of the convergence-containment method can nevertheless be applied (Panet 1986) (see also Muir-Wood 1975; Einstein and Schwartz 1979).

If at a distance $d$ from the face corresponding to a value $\lambda_d$ of the deconfinement ratio, an elastic support with normal stiffness $K_{sn}$ and bending stiffness $K_{sf}$ is installed, the stresses $\sigma_R$ and $\tau_{r\theta}$ at equilibrium are given by:

$$\sigma_R = p + q\cos\theta$$
$$\tau_{R\theta} = t\sin 2\theta \tag{3.26}$$

where $p$, $q$ and $t$ depend on the conditions of contact between the support and the ground.

If we consider that the support has no bending stiffness, as in the case of a support by bolting, or that it can be neglected as compared to the normal stiffness, $p$, $q$ and $t$ can obtained as:

$$p = \frac{1}{2}\frac{k_{sn}}{1 + k_{sn}}(1 - \lambda_d)(1 + K_0)\sigma_0$$
$$q = -\frac{1}{2}\frac{3 - 4\nu}{1 - 2\nu}(1 - \lambda_d)(1 - K_0)\sigma_0 \tag{3.27}$$
$$t = 0$$

which gives:

$$\sigma_R = p_s = \frac{1}{2}\left[\frac{k_{sn}}{1+k_{sn}}(1+K_0) - \frac{3-4v}{1-2v}(1-K_0)\cos 2\theta\right](1-\lambda_d)\sigma_0 \quad (3.28)$$

If the bending stiffness cannot be neglected, two extreme contact conditions can be considered: perfect adherence between the lining and the ground (tied contact) or zero friction at the contact (perfect slip).

Assuming tied contact between the lining and the ground implies the continuity of the radial and the hoop displacements at the interface between the ground and the support. Under this condition, $p, q, t$ are given by:

$$\left(1+k_{sn}+k_{sf}\right)p = \frac{1}{2}\left(k_{sn}+k_{sf}\right)(1-\lambda_d)(1+K_0)\sigma_0$$

$$(1-2v)k_{sn}q - \frac{1}{2}[1+4(1-v)k_{sn}]t = -\frac{1}{2}k_{sn}(3-4v)(1-\lambda_d)(1-K_0)\sigma_0$$

$$2[1+3(5-6v)k_{sf}]q - [1+6(4-6v)k_{sf}]t = -9k_{sf}(3-4v)(1-\lambda_d)(1-K_0)\sigma_0$$

where:

$$k_{sn} = \frac{K_{sn}}{2G}; k_{sf} = \frac{K_{sf}}{2G}$$

$$(3.29)$$

Under the assumption of perfect slip, only the radial displacement at the interface between the ground and the support is continuous and, moreover $\tau_{R\theta}$ is zero; We obtain:

$$\left(1+k_{sn}+k_{sf}\right)p = \frac{1}{2}\left(k_{sn}+k_{sf}\right)(1-\lambda_d)(1+K_0)\sigma_0$$

$$2[1+3(5-6v)k_{sf}]q = -9k_{sf}(3-4v)(1-\lambda_d)(1-K_0)\sigma_0 \qquad (3.30)$$

$$t = 0$$

By noting that in the previous equations $p, q$ and $t$ are proportional to $(1-\lambda_d)$, $p$ is proportional to $(1+K_0)\sigma_0$ and $q$ and $t$ are proportional to $(1-K_0)\sigma_0$, we can rewrite the previous equations as follows:

$$p = \frac{1}{2}\alpha_1(1-\lambda_d)(1+K_0)\sigma_0$$

$$q = \frac{1}{2}\alpha_2(1-\lambda_d)(1-K_0)\sigma_0 \qquad (3.31)$$

$$t = \frac{1}{2}\beta_2(1-\lambda_d)(1-K_0)\sigma_0$$

with:

$$\alpha_1 = \frac{k_{sn} + k_{sf}}{1 + k_{sn} + k_{sf}} \tag{3.32}$$

For tied contact, $\alpha_2$ and $\beta_2$ are given by:

$$(1 - 2\nu)k_{sn}\alpha_2 - \frac{1}{2}[1 + 4(1 - \nu)k_{sn}]\beta_2 = -(3 - 4\nu)k_{sn}$$

$$[1 + 3(5 - 6\nu)k_{sf}]\alpha_2 - \frac{1}{2}[1 + 6(4 - 6\nu)k_{sf}]\beta_2 = -9(3 - 4\nu)k_{sf} \tag{3.33}$$

and for perfect slip:

$$\alpha_2 = -\frac{9(3 - 4\nu)k_{sf}}{1 + 3(5 - 6\nu)k_{sf}}$$

$$\beta_2 = 0 \tag{3.34}$$

The coefficients $\alpha_1, \alpha_2, \beta_2$ only depend on $k_{sn}$ and $k_{sf}$ which characterize the relative support stiffness to the ground stiffness.

These expressions are used to determine the normal force $N$ and the bending moment $M$ in the support:

$$N = \frac{1}{2}[n_1(1 + K_0) + n_2(1 - K_0)\cos 2\theta](1 - \lambda_d)\sigma_0 R$$

$$M = \frac{1}{2}m_2(1 - K_0)\cos 2\theta(1 - \lambda_d)\sigma_0 R$$

where:

$$n_1 = \alpha_1$$

$$n_2 = \frac{1}{3}(2\beta_2 - \alpha_2)$$

$$m_2 = \frac{1}{6}(\beta_2 - 2\alpha_2) \tag{3.35}$$

For example, for a support such that $K_{sn} = 2 \times 10^3 K_{sf}$, the values of the coefficients $n_1, n_2, m_2$ are given in Table 3.1 for the two cases of perfect adherence and perfect slip for various values of the relative stiffness $k_{sn}$ (Poisson's ratio of the ground is taken equal to 0.3).

At equilibrium, the displacements $u_R$ and $v_R$ at the extrados of the support are given by:

$$\frac{u_R}{R} = U_1 + U_2 \cos 2\theta$$

$$\frac{v_R}{R} = V_2 \sin 2\theta \tag{3.36}$$

**Table 3.1** Coefficients $n_1, n_2, m_2$ for the evaluation of the forces acting in the support

| $k_{sn}$ | | 0.25 | 0.50 | 0.75 | 1 | 2 | 5 | 10 |
|---|---|---|---|---|---|---|---|---|
| $n_1$ | | 0.20 | 0.33 | 0.43 | 0.50 | 0.67 | 0.83 | 0.91 |
| $n_2$ | 1 | 0.28 | 0.41 | 0.48 | 0.53 | 0.62 | 0.70 | 0.73 |
| | 2 | $6.74 \times 10^{-4}$ | $1.35 \times 10^{-3}$ | $2.02 \times 10^{-3}$ | $2.69 \times 10^{-3}$ | $5.35 \times 10^{-3}$ | $1.32 \times 10^{-3}$ | $2.58 \times 10^{-3}$ |
| $m_2$ | 1 | $6.32 \times 10^{-4}$ | $1.22 \times 10^{-3}$ | $1.80 \times 10^{-3}$ | $2.37 \times 10^{-3}$ | $4.61 \times 10^{-3}$ | $1.12 \times 10^{-2}$ | $2.17 \times 10^{-2}$ |
| | 2 | $6.74 \times 10^{-4}$ | $1.35 \times 10^{-3}$ | $2.02 \times 10^{-3}$ | $2.69 \times 10^{-3}$ | $5.35 \times 10^{-3}$ | $1.32 \times 10^{-2}$ | $2.58 \times 10^{-2}$ |

1. Perfect adherence
2. Perfect slip

where $U_1, U_2$ and $V_2$ are linked to $p$, $q$ and $t$ by the following relationships:

$$p = (K_{sn} + K_{sf})U_1$$
$$q = K_{sn}(U_2 + 2V_2) + 9K_{sf}U_2 \qquad (3.37)$$
$$t = 2K_{sn}(U_2 + 2V_2)$$

The corresponding displacements in the ground are obtained by adding the displacements that occurred before the installation of the support.

## Appendix 1: Complex Variable Method for Solving Two-Dimensional Elasticity Problems

Two-dimensional problems in elasticity can be solved using the complex variable method. In this method, the position of a material point having coordinates $(x, y)$ is represented by the complex number $z = x + iy$ and the displacements and stresses are represented in terms of two analytic functions of the complex variable $z$.

For plane stress and plane stress isotropic elasticity, the stress–strain relationships can be written as:

$$8G\varepsilon_x = (\kappa + 1)\sigma_x + (\kappa - 3)\sigma_y$$
$$8G\varepsilon_y = (\kappa - 3)\sigma_x + (\kappa + 1)\sigma_y \qquad (3.38)$$
$$\tau_{xy} = 2G\varepsilon_{xy}$$

where

$$\kappa = 3 - 4\nu \text{ for plane strain,}$$

$$\kappa = \frac{3 - \nu}{1 + \nu} \text{ for plane stress} \tag{3.39}$$

In absence of body forces, the stresses are given by the Airy stress function:

$$\sigma_x = \frac{\partial^2 U}{\partial y^2}, \quad \sigma_y = \frac{\partial^2 U}{\partial x^2}, \quad \tau_{xy} = -\frac{\partial^2 U}{\partial x \partial y} \tag{3.40}$$

and one can easily verify that the equilibrium equations

$$\frac{\partial \sigma_x}{\partial x} + \frac{\partial \tau_{xy}}{\partial y} = 0$$

$$\frac{\partial \tau_{xy}}{\partial x} + \frac{\partial \sigma_y}{\partial y} = 0 \tag{3.41}$$

are automatically fulfilled.

In order for the strain compatibility equation $\frac{\partial^2 \varepsilon_x}{\partial y^2} + \frac{\partial^2 \varepsilon_y}{\partial x^2} = 2\frac{\partial^2 \varepsilon_{xy}}{\partial x \partial y}$ to be verified, the Airy stress function $U$ must satisfy the biharmonic equation:

$$\nabla^2 (\nabla^2 U) = \frac{\partial^4 U}{\partial x^4} + 2\frac{\partial^4 U}{\partial x^2 \partial y^2} + \frac{\partial^4 U}{\partial y^4} = 0 \tag{3.42}$$

Thus the Airy stress function $U$ can be expressed in terms of two analytic functions $\phi(z)$ and $\chi(z)$:

$$U = \text{Re}\{\bar{z}\phi(z) + \chi(z)\} \tag{3.43}$$

On can deduce the expressions of the stresses and displacements:

$$\sigma_x + \sigma_y = 4\text{Re}\{\phi'(z)\}$$
$$\sigma_y - \sigma_x + 2i\tau_{xy} = 2[\bar{z}\phi''(z) + \psi'(z)]$$
$$\psi(z) = \chi'(z)$$
$$2G(u_x + iu_y) = \kappa\phi(z) - z\overline{\phi'(z)} - \overline{\psi(z)} \tag{3.44}$$

or in polar coordinates as:

$$\sigma_r + \sigma_\theta = \sigma_x + \sigma_y = 4\text{Re}\{\phi'(z)\}$$

$$\sigma_\theta - \sigma_r + 2i\tau_{r\theta} = \left(\sigma_y - \sigma_x + 2i\tau_{xy}\right)e^{2i\theta}$$

$$= 2\left[\overline{z}\phi''(z) + \psi'(z)\right]e^{2i\theta} \tag{3.45}$$

$$u_r + iu_\theta = \left(u_x + iu_y\right)e^{-i\theta} = \frac{1}{2G}\left(\kappa\phi(z) - z\overline{\phi'(z)} - \overline{\psi(z)}\right)e^{-i\theta}$$

Solving a two-dimensional elasticity problem thus consists in finding the two potential functions $\phi(z)$ and $\psi(z)$.

For an excavation of circular section of radius $R$ under a far field stress $\sigma_1$ along the $x$-axis and $\sigma_2$ along the $y$-axis (Fig. 3.10), the potential functions are of the form:

$$\phi(z) = Az + \frac{B}{z}$$

$$\psi(z) = A'z + \frac{B'}{z} + \frac{C'}{z^3} \tag{3.46}$$

where $A$, $B$, $A'$, $B'$ and $C'$ are constants that can be calculated using the boundary conditions. For stress free inner boundaries, the following solution is obtained:

$$\sigma_r = \frac{1}{2}(\sigma_1 + \sigma_2)\left(1 - \frac{R^2}{r^2}\right) + \frac{1}{2}(\sigma_1 - \sigma_2)\left(1 - \frac{4R^2}{r^2} + \frac{3R^4}{r^4}\right)\cos 2\theta$$

$$\sigma_\theta = \frac{1}{2}(\sigma_1 + \sigma_2)\left(1 + \frac{R^2}{r^2}\right) - \frac{1}{2}(\sigma_1 - \sigma_2)\left(1 + \frac{3R^4}{r^4}\right)\cos 2\theta$$

$$\tau_{r\theta} = -\frac{1}{2}(\sigma_1 - \sigma_2)\left(1 + \frac{2R^2}{r^2} - \frac{3R^4}{r^4}\right)\sin 2\theta$$

$$\frac{8G}{R}u_r = (\sigma_1 + \sigma_2)\left((\kappa - 1)\frac{r}{R} + \frac{2R}{r}\right) \tag{3.47}$$

$$+(\sigma_1 - \sigma_2)\left\{2\frac{r}{R} + \frac{2R}{r}\left(\kappa + 1 - \frac{R^2}{r^2}\right)\right\}\cos 2\theta$$

$$\frac{8G}{R}u_\theta = (\sigma_1 - \sigma_2)\left[-\frac{2r}{R} + \frac{2R}{r}\left\{1 - \kappa - \frac{R^2}{r^2}\right\}\right]\sin 2\theta$$

## Appendix 2: Stress and Displacement Fields Around a Tunnel with Circular Cross Section Excavated in an Elastic Medium with Transverse Isotropy

We consider a circular tunnel excavated in an infinite linear elastic rock mass with transverse isotropy. This type of anisotropy is commonly encountered in sedimentary

**Fig. 3.10** Tunnel with
circular cross section under
anisotropic far field stress
state

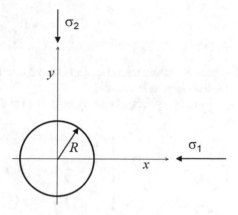

and metamorphic foliated rock. The most unfavorable situation corresponds to the
case when the axis of the tunnel is parallel to the plane of isotropy safety. Plane problems for anisotropic elastic bodies have been largely studied in the works of Green
and Taylor (1939, 1945a, b), Green and Zerna (1968) and by Lekhnitskii (1963).
Hefny and Lo (1999) have applied Green's theory for unlined circular tunnels excavated in an elastic transversely isotropic medium. A summary of the mathematical
derivations can also be found in Manh Huy Tran's Ph.D. thesis (2014).

The global coordinate system is defined in such a way that the $z$-axis is the tunnel
axis, the $x$-axis lies in the plane of isotropy and the $y$-axis is normal to it (Fig. 3.11).
It is assumed that the plane of isotropy makes an angle $\beta$ with the horizontal plane.

The initial state of stress is assumed to be homogeneous and anisotropic. In the
plane of a tunnel cross section, it is expressed as:

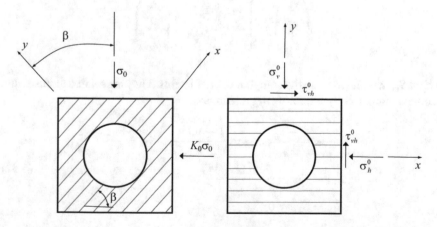

**Fig. 3.11** Tunnel with circular cross section excavated in a transverse isotropic medium

$$\boldsymbol{\sigma} = \begin{pmatrix} \sigma_0 & 0 \\ 0 & K_0\sigma_0 \end{pmatrix} \tag{3.48}$$

where $K_0$ is the lateral earth pressure coefficient (ratio between the horizontal and vertical far-field stresses).

In the $(x, y)$ coordinate system, the far-field stresses are given by:

$$\begin{cases} \sigma_v^0 = \dfrac{\sigma_0}{2}((1 + K_0) + (1 - K_0)\cos 2\beta) \\ \sigma_h^0 = \dfrac{\sigma_0}{2}((1 + K_0) - (1 - K_0)\cos 2\beta) \\ \tau_{vh}^0 = \dfrac{\sigma_0}{2}(1 - K_0)\sin 2\beta \end{cases} \tag{3.49}$$

The excavation process is simulated by reducing the stresses at the inner wall of the tunnel:

$$\begin{cases} \Delta\sigma_v^0 = -\lambda\sigma_v^0 = -\dfrac{\lambda\sigma_0}{2}((1 + K_0) + (1 - K_0)\cos 2\beta) \\ \Delta\sigma_h^0 = -\lambda\sigma_h^0 = -\dfrac{\lambda\sigma_0}{2}((1 + K_0) - (1 - K_0)\cos 2\beta) \\ \Delta\tau_{vh}^0 = -\lambda\tau_{vh}^0 = -\dfrac{\lambda\sigma_0}{2}(1 - K_0)\sin 2\beta \end{cases} \tag{3.50}$$

where $\lambda$ is the deconfinement rate.

Elastic constitutive relationships for plane strain are written as:

$$\begin{pmatrix} \varepsilon_x \\ \varepsilon_y \\ \varepsilon_{xy} \end{pmatrix} = \begin{bmatrix} S_{11} & S_{12} & 0 \\ S_{12} & S_{22} & 0 \\ 0 & 0 & S_{33} \end{bmatrix} \begin{pmatrix} \sigma_x \\ \sigma_y \\ \tau_{xy} \end{pmatrix} \tag{3.51}$$

where $S_{11}, S_{12}, S_{22}, S_{33}$ are the compliance coefficients. They are related to the elastic ground parameters by the following relations:

$$\begin{aligned} & S_{11} = \dfrac{1 - \nu_h^2}{E_h}, \quad S_{22} = \dfrac{1 - \nu_{vh}\nu_{hv}}{E_v}, \quad S_{33} = \dfrac{1}{G_{vh}} \\ & S_{12} = S_{21} = \dfrac{-\nu_{vh}(1 + \nu_h)}{E_v}, \\ & S_{13} = S_{31} = 0 \end{aligned} \tag{3.52}$$

where $E_h$, $E_v$ are the Young's modulus in the plane of isotropy and in the direction normal to it respectively; $\nu_h$ is the Poisson's ratio in the plane of isotropy, $\nu_{hv}$ et $\nu_{vh}$ are the Poisson's coefficients for a load in the plane of isotropy and normal to it respectively, $(E_h/\nu_{hv} = E_v/\nu_{vh})$; $G_{vh}$ is the elastic shear modulus in any plane normal to the plane of isotropy.

Stresses are derived from the Airy stress function $\underline{U}$ (Eq. 3.41). Using the constitutive Eq. (3.51) and the strain compatibility equation for two-dimensional problems $2\frac{\partial^2 \varepsilon_{xy}}{\partial x \partial y} = \frac{\partial^2 \varepsilon_x}{\partial y^2} + \frac{\partial^2 \varepsilon_y}{\partial x^2}$ yields the following differential equation for the Airy stress function (Jaeger et Cook 1976):

$$S_{11}\frac{\partial^4 U}{\partial y^2} + (2S_{12} + S_{33})\frac{\partial^4 U}{\partial x^2 \partial y^2} + S_{22}\frac{\partial^4 U}{\partial x^2} = 0 \qquad (3.53)$$

which can be written as:

$$\left(\frac{\partial^2}{\partial x^2} + \alpha_1 \frac{\partial^2}{\partial y^2}\right)\left(\frac{\partial^2}{\partial x^2} + \alpha_2 \frac{\partial^2}{\partial y^2}\right)U = 0 \qquad (3.54)$$

with

$$\alpha_1 \alpha_2 = \frac{S_{11}}{S_{22}} \text{ and } \alpha_1 + \alpha_2 = \frac{S_{33} + 2S_{12}}{S_{22}} \qquad (3.55)$$

We have seen in Appendix 1 that two-dimensional problems in elasticity are commonly solved using the complex variable method. In this method, the displacements and stresses field are represented in terms of analytic functions of a complex variable $z = x + iy$. In the case of transversely isotropy, following Green and Taylor (1939, 1945a, b), two additional complex variables $z_1$ and $z_2$ are introduced:

$$\text{for } k = 1, 2 \; z_k = (z + \gamma_k \overline{z}), \text{ with } \gamma_k = \frac{\sqrt{\alpha_k} - 1}{\sqrt{\alpha_k} + 1}; \; |\gamma_k| < 1 \qquad (3.56)$$

The general solution of Eq. (3.54) for the Airy function can be expressed in terms of two analytic functions ($\Omega_1$, $\Omega_2$) and their conjugates:

$$U = \sum_{k=1}^{2}\left(\Omega_k(z_k) + \overline{\Omega_k(z_k)}\right) \qquad (3.57)$$

The stress and displacement fields are expressed as:

**Stresses**:

$$\sigma_{xx} = -\sum_{k=1}^{2}\left((\gamma_k - 1)^2\Omega_k''(z_k) + \left(\overline{\gamma}_k - 1\right)^2\overline{\Omega_k''(z_k)}\right)$$

$$\sigma_{yy} = \sum_{k=1}^{2}\left((\gamma_k + 1)^2\Omega_k''(z_k) + \left(\overline{\gamma}_k + 1\right)^2\overline{\Omega_k''(z_k)}\right) \qquad (3.58)$$

$$\tau_{xy} = -\frac{1}{i}\sum_{k=1}^{2}\left((\gamma_k^2 - 1)\Omega_k''(z_k) - \left(\overline{\gamma}_k^2 - 1\right)\overline{\Omega_k''(z_k)}\right)$$

**Displacements**:

$$u = \frac{1}{2}\sum_{k=1}^{2}\left[(\delta_k + \rho_k)\Omega_k'(z_k) + \left(\overline{\delta}_k + \overline{\rho_k}\right)\Omega_k'(\overline{z}_k)\right]$$

$$v = \frac{1}{2i}\sum_{k=1}^{2}\left[(\delta_k - \rho_k)\Omega_k'(z_k) - \left(\overline{\delta}_k - \overline{\rho_k}\right)\Omega_k'(\overline{z}_k)\right] \qquad (3.59)$$

with

$$\delta_1 = (1 + \gamma_1)\beta_2 - (1 - \gamma_1)\beta_1; \quad \delta_2 = (1 + \gamma_2)\beta_1 - (1 - \gamma_2)\beta_2$$
$$\rho_1 = (1 + \gamma_1)\beta_2 + (1 - \gamma_1)\beta_1; \quad \rho_2 = (1 + \gamma_2)\beta_1 + (1 - \gamma_2)\beta_2$$
$$\beta_1 = S_{12} - S_{22}\alpha_1 \; ; \; \beta_2 = S_{12} - S_{22}\alpha_2$$

**Boundary conditions**:

Boundary conditions are expressed in terms of the resultant tractions acting $(X, Y)$ on the boundary:

$$P = X + iY = 2i\frac{\partial U}{\partial \overline{z}} = 2i\sum_{k=1}^{2}\left(\gamma_k\Omega_k'(z_k) + \overline{\Omega_k'(z_k)}\right) \qquad (3.60)$$

For practical applications, the potential functions $\Omega_1(z_1)$ and $\Omega_2(z_2)$ are searched in the form of infinite power series. For the problem considered for which the medium extends to infinity, only the negative powers of the complex variable should be included to ensure finite stress at infinity. The first derivative of the potential functions can be assumed to be of the form:

$$\Omega_k'(z_k) = \sum_{n=0}^{\infty} A_{kn}(z_k)^{-2n-1} \qquad (3.61)$$

where the coefficients $A_{kn}$ are evaluated from the boundary conditions (3.60).

In polar coordinates,

$$\Omega'_k(z_1) = \frac{1}{r} \sum_{s=0}^{\infty} f_{ks}(r) e^{-i(2s+1)\theta}$$

with $f_{k0}(r) = A_{k0}$

and for $s \geq 1$, $f_{ks}(r) = A_{k0}(\gamma_k)^s(-1)^s + \sum_{n=1}^{s} \frac{A_{kn}}{r^{2n}} \binom{n+s}{s-n} (\gamma_k)^{s-n}(-1)^{s-n}$

$$(3.62)$$

At the tunnel wall, $r = R$:

$$P(\theta) = -\int_A^P (\sigma_r + i\tau_{r\theta}) \frac{dz}{d\theta} d\theta = \frac{\lambda\sigma_0}{2} R \int_0^\theta \left(e^{-2i\theta}(K_0-1)e^{-2i\beta} + (K_0+1)\right)e^{i\theta} d\theta$$

$$= -i\frac{\lambda\sigma_0}{2} R\left((1-K_0)e^{-2i\beta}e^{-i\theta} + (1+K_0)e^{i\theta}\right)$$

$$(3.63)$$

Using Eqs. (3.60) and (3.61), the resultant tractions acting on the boundary can be written as:

$$P(\theta) = \frac{2i}{R} \sum_{k=1}^{2} \left(\gamma_k f_{ks}(R) e^{-i(2s+1)\theta} + \overline{f_{ks}}(R) e^{i(2s+1)\theta}\right) \qquad (3.64)$$

By identification of the terms in $e^{-i(2s+1)\theta}$ and $e^{i(2s+1)\theta}$ we obtain:

$$\sum_{k=1}^{2} \gamma_k f_{k0}(R) = -\frac{\lambda\sigma_0}{4} R^2\left[(1-K_0)e^{-2i\beta}\right]$$

$$\sum_{k=1}^{2} \overline{f}_{k0}(R) = -\frac{\lambda\sigma_0}{4} R^2\left[(1+K_0)\right]$$

$$(3.65)$$

and for $s \geq 1$, $\quad \begin{array}{c} \sum_{k=1}^{2} \gamma_k f_{ks}(R) = 0 \\ \sum_{k=1}^{2} \overline{f}_{ks}(R) = 0 \end{array}$

The above equations are fulfilled for:

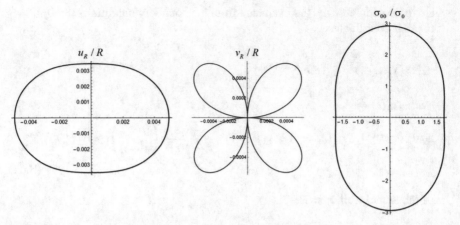

**Fig. 3.12** Polar diagrams of the radial and hoop displacements and of the hoop stress at the wall of fully deconfined tunnel ($\lambda = 1$) (the ground elastic parameters and the initial stress state correspond to the conditions of the Meuse/Haute-Marne Underground Laboratory of Andra)

$$A_{10} = \frac{\lambda\sigma_0 R^2}{4(\gamma_2-\gamma_1)}\left[(1-K_0)e^{-2i\beta} - \gamma_2(1+K_0)\right]$$
$$A_{20} = \frac{\lambda\sigma_0 R^2}{4(\gamma_1-\gamma_2)}\left[(1-K_0)e^{-2i\beta} - \gamma_1(1+K_0)\right]$$
$$\text{and for } n \geq 1, \quad A_{kn} = A_{k0}R^{2n}\gamma_k^n \frac{2n!}{n!(n+1)!}$$

$$(3.66)$$

As for example, the polar diagrams of the radial and hoop displacements and of the hoop stress at the wall of fully deconfined tunnel ($\lambda = 1$) are plotted in Fig. 3.12. In this example, the ground elastic parameters and the initial stress state correspond to the conditions of the Meuse/Haute-Marne Underground Laboratory of Andra which is situated at a depth of 490 m and for which the isotropic plane is horizontal:

$E_h = 5000$ MPa; $E_v = 4000$ MPa; $G_{vh} = 1630$ MPa; $\nu_h = 0.24$; $\nu_{vh} = 0.33$;
$\sigma_0 = 12$ MPa; $K_0 = 1.3$; $\beta = 0$

This solution can be extended to the case of an excavation with arbitrary cross section shape by using the method of conformal mapping as proposed by Tran-Manh et al. (2015).

## References

Green AE, Taylor GI (1939) Stress systems in aeolotropic plates I. Proc R Soc Ser A Math Phys Sci 173(953):162–172
Green AE, Taylor GI (1945a) Stress systems in aeolotropic plates III. Proc R Soc Ser A Math Phys Sci 184:181–195

Green AE, Taylor GI (1945b) Stress systems in aeolotropic plates VI. Proc R Soc Ser A Math Phys Sci 184:289–300

Green AE, Zerna W (1968) Theoretical elasticity. Dover Publications, Inc.

Einstein HH, Schwartz CW (1979) Simplified analysis for tunnel supports. J Geotech Eng Div ASCE 105(4):499–518

Hefny AM, Lo KY (1999) Analytical solutions for stresses and displacements around tunnels driven in cross-anisotropic rocks. Int J Numer Anal Methods Geomech 23:161–177

Jaeger JC, Cook NGW (1976) Fundamentals of rock mechanics, New York

Lekhnitskii SG (1963) Theory of elasticity of an anisotropic elastic body. Holden-Day Inc., San Francisco

Muir-Wood AM (1975) The circular tunnel in elastic ground. Géotechnique 25(1):115–127

Panet M (1986) Calcul du soutènement des tunnels à section circulaire par la méthode convergence-confinement avec un champ de contraintes initiales anisotrope. Tunnels Et Ouvrages Souterrains 77:228–232

Tran Manh H (2014) Comportement des tunnels en terrain poussant, Thèse de doctorat de l'Université Paris-Est

Tran Manh H, Sulem J, Subrin D (2015) A closed-form solution for tunnels with arbitrary cross section excavated in elastic anisotropic ground. Rock Mech Rock Eng 48(1):277–288

# Chapter 4
# The Convergence-Confinement Method for a Tunnel Driven in an Elasto-Plastic Medium

We saw in Chap. 3 that for an elastic behavior of the ground, the shear stress $\sigma_\theta - \sigma_r$ is maximum at the wall of the tunnel. Therefore, shear failure first appears at the wall and then progresses inside the ground. A *plastic zone*, also called *yielded zone*, develops around the excavation. The analysis of the conditions for the onset of this plastic zone and its extension is a key point for the choice of the excavation method and the type of support.

Under the assumption of axi-symmetry, corresponding to an isotropic behavior and isotropic initial state of stress, and for commonly used plasticity models such as Tresca, Mohr–Coulomb and Hoek and Brown, closed-form solutions can be obtained for the stress and deformation fields (Panet 1993). These solutions are presented in this chapter in order to establish the *Ground Reaction Curve* (GRC) or convergence curve of the ground.

## 4.1 Usual Plasticity Models

The most commonly used plasticity models for underground applications are the Tresca, Mohr–Coulomb, and Hoek and Brown models. They are formulated from a yield criterion $F(\sigma)$ which defines the boundary of the elastic domain and a plastic potential $Q(\sigma)$ from which the flow rule is derived. For associated plasticity, the yield criterion and the plastic potential are identical (normality flow rule is fulfilled), whereas for non-associated models, these two functions are different. Normality flow rule generally leads to excessive dilatancy which is not verified in experimental observations. A non-associated flow rule is then considered ($Q \neq F$) with a dilatancy angle less than the internal friction angle. The assumption of plastic deformation at constant volume corresponds to a zero dilatancy angle.

**Fig. 4.1** Tresca criterion

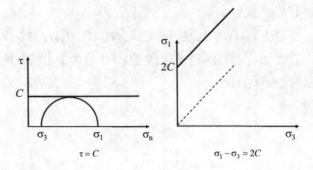

$$\tau = C \qquad\qquad \sigma_1 - \sigma_3 = 2C$$

### 4.1.1  Tresca Yield Criterion

The Tresca yield criterion (Fig. 4.1) is suited for purely cohesive materials. It is expressed in terms of the two extreme principal stresses, $\sigma_1$, major principal stress and $\sigma_3$, minor principal stress. The only strength parameter is the cohesion $C$:

$$F(\sigma_1, \sigma_3) = \sigma_1 - \sigma_3 - 2C = 0 \tag{4.1}$$

### 4.1.2  Mohr–Coulomb Yield Criterion

Mohr–Coulomb criterion (Fig. 4.2) introduces an internal friction angle $\phi$ in addition to the cohesion $C$:

$$F(\sigma_1, \sigma_3) = \sigma_1 - K_p \sigma_3 - \sigma_c = 0 \tag{4.2}$$

where

$$K_p = \frac{1 + \sin \phi}{1 - \sin \phi} = \tan^2 \left( \frac{\pi}{4} + \frac{\phi}{2} \right)$$

**Fig. 4.2** Mohr Coulomb criterion

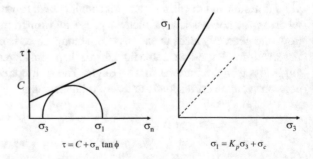

$$\tau = C + \sigma_n \tan \phi \qquad\qquad \sigma_1 = K_p \sigma_3 + \sigma_c$$

$$\sigma_c = \frac{2C\cos\phi}{1 - \sin\phi} \text{ (uniaxial strength)} \tag{4.3}$$

We can also introduce the parameter $H = \frac{C}{\tan\phi}$ and write the Mohr–Coulomb criterion in the following form:

$$F(\sigma_1, \sigma_3) = \sigma_1 - H - K_p(\sigma_3 - H) = 0 \tag{4.4}$$

### 4.1.3 Hoek and Brown Yield Criterion

For rock masses, the parabolic Hoek and Brown (1980) yield criterion is commonly used (Fig. 4.3). It is written as:

$$\sigma_1 = \sigma_3 + \sigma_c\sqrt{m\frac{\sigma_3}{\sigma_c} + s} \tag{4.5}$$

where $m$ and $s$ are two parameters that can be evaluated from triaxial tests on samples of intact rock and from geotechnical classifications of the rock mass such as the *Rock Mass Rating* (RMR) classification:

$$m = m_r \exp\left(\frac{RMR - 100}{14 I_m}\right)$$
$$s = \exp\left(\frac{RMR - 100}{6 I_s}\right) \tag{4.6}$$

**Fig. 4.3** Linear Mohr–Coulomb criterion ($K_p = 3$) and parabolic Hoek and Brown criterion ($m = 7, \quad s = 1$)

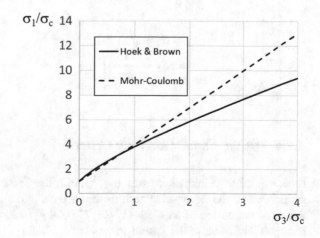

where:

$m_r$ is the value of parameter $m$ corresponding to a sound unfractured rock (RMR = 100);

$I_m = 1$ and $I_s = 1$ for an undisturbed rock mass;

$I_m = 2$ et $I_s = 1.5$ for a disturbed rock mass.

A generalized Hoek and Brown criterion has been later proposed (Hoek et al. 2002):

$$\sigma_1 = \sigma_3 + \sigma_c \left( m \frac{\sigma_3}{\sigma_c} + s \right)^a \tag{4.7}$$

and empirical relationships have been proposed to evaluate parameters $m$, $s$ and $a$, from the value of the Geological Strength Index (GSI) of the rock mass:

$$m = m_r \exp\left( \frac{GSI - 100}{28 - 14D} \right)$$

$$s = \exp\left( \frac{GSI - 100}{9 - 3D} \right)$$

$$a = \frac{1}{2} + \frac{1}{6} \left( e^{-GSI/15} - e^{-20/3} \right) \tag{4.8}$$

In the above expressions, $D$ is a factor which depends upon the degree of disturbance to which the rock mass has been subjected by blast damage and stress relaxation. It varies from 0 for undisturbed in situ rock masses to 1 for very disturbed rock masses. Guidelines for the selection of $D$ are given in the paper of Hoek et al. (2002).

Londe (1988) has proposed a remarkably simple dimensionless expression for the Hoek and Brown yield criterion which is useful in the development of closed-form solutions:

$$S_1 = S_3 + \sqrt{S_3}$$

$$\text{with: } S_1 = \frac{\sigma_1}{m\sigma_c} + \frac{s}{m^2}; \quad S_3 = \frac{\sigma_3}{m\sigma_c} + \frac{s}{m^2} \tag{4.9}$$

### 4.1.4 Other Usual Plasticity Criteria

In the space of principal stresses $(\sigma_1, \sigma_2, \sigma_3)$, the Tresca criterion is represented by prism and the Mohr–Coulomb criterion by a pyramid; for these two criteria, the direction cosine of axis are $(1/\sqrt{3}, /1/\sqrt{3}, /1/\sqrt{3})$ and the base is a regular hexagon. The edges of the Tresca prism or of the Mohr–Coulomb pyramid generates particular difficulties in numerical codes. Thus, a smooth form derived from these

**Fig. 4.4** Traces of Mohr–Coulomb and Drucker-Prager yield surfaces in the deviatoric plane. ($s_1$, $s_2$, $s_3$ are the principal deviatoric stresses)

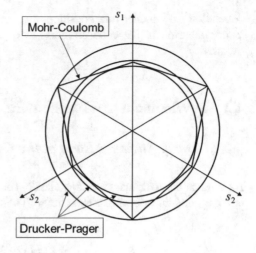

criteria is commonly used by considering the circle inscribed or circumscribed to these hexagons (Fig. 4.4). These criteria involve the first two invariants of the stress tensors:

$$I_1 = \sigma_1 + \sigma_2 + \sigma_3$$
$$J_2 = \frac{1}{6}\left[(\sigma_1 - \sigma_2)^2 + (\sigma_2 - \sigma_3)^2 + (\sigma_3 - \sigma_1)^2\right] \qquad (4.10)$$

- Von Mises criterion:

$$\sqrt{3J_2} - k = 0$$
$$\text{with}: k = \sqrt{3}C \qquad (4.11)$$
$$\text{or}: k = 2C$$

depending on whether we consider the cylinder inscribed or circumscribed to the Tresca prism.

Drucker-Prager criterion:

$$\alpha I_1 + \sqrt{3J_2} - k = 0$$
$$\text{with}: \alpha = \frac{\tan\phi}{\sqrt{3 + 4\tan^2\phi}}$$
$$\text{or}: \alpha = \frac{2\sin\phi}{3 + \sin\phi}$$
$$\text{or}: \alpha = \frac{2\sin\phi}{3 - \sin\phi} \qquad (4.12)$$

depending on whether one considers the inscribed cone, the cone circumscribed in compression or the cone circumscribed in extension to the Mohr–Coulomb's pyramid.

## 4.2  Development of a Plastic Zone

### 4.2.1  Deconfinement Ratio at the Onset of Plasticity

As long as the rock mass remains elastic, for a rate of deconfinement $\lambda$, the radial stress $\sigma_R$ and the hoop stress $\sigma_\theta$ at the wall of a tunnel with circular section are given by:

$$\sigma_R = (1 - \lambda)\sigma_0$$
$$\sigma_\theta = (1 + \lambda)\sigma_0 \tag{4.13}$$

The yield criterion is first reached at the wall for a deconfinement ratio $\lambda_e$ such as (Fig. 4.5):

$$F((1 + \lambda_e)\sigma_0, (1 - \lambda_e)\sigma_0) = 0 \tag{4.14}$$

For Tresca criterion:

$$\lambda_e = \frac{C}{\sigma_0} \tag{4.15}$$

For Mohr–Coulomb criterion:

**Fig. 4.5**  Yield criterion and elasto-plastic stress path

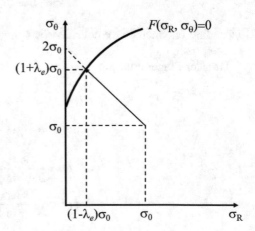

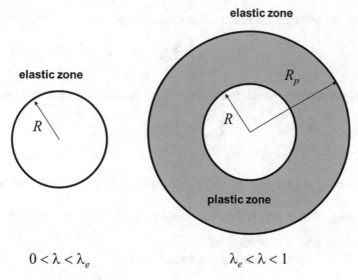

elastic zone

elastic zone

$R_p$

$R$

$R$

plastic zone

$0 < \lambda < \lambda_e$ $\lambda_e < \lambda < 1$

**Fig. 4.6** Onset of a plastic zone for $\lambda > \lambda_e$

$$\lambda_e = \frac{1}{K_p + 1}\left(K_p - 1 + \frac{\sigma_c}{\sigma_0}\right) \tag{4.16}$$

For Hoek and Brown criterion:

$$\lambda_e = \frac{\sigma_c}{8\sigma_0}\left[\sqrt{m^2 + 16m\frac{\sigma_0}{\sigma_c} + 16s} - m\right] \tag{4.17}$$

In the case of an unsupported tunnel, no plastic zone appears if the hoop stress at the wall verifies $\sigma_\theta < \sigma_c$, which corresponds to $\lambda_e > 1$.

When the deconfinement ratio becomes greater than $\lambda_e$, a plastic zone of radius $R_p$ develops around the tunnel (Fig. 4.6). The radius of the plastic zone increases with the deconfinement ratio. In the plastic zone, the radial stress increases from $(1 - \lambda)\sigma_0$ at the tunnel wall up to $(1 - \lambda_e)\sigma_0$ at the boundary between the plastic zone and the elastic zone. At this point, the hoop stress passes through a maximum which is an angular point.

### 4.2.2 Stability Number

A brief assessment of the development of this plastic zone can be made using the notion of the stability number $N$ which was first introduced by Broms and Benner-mark (1967) to analyze the short-term stability of the tunnel face for an excavation at a depth $H$ in a clay soil of undrained cohesion $C_u$, $N = \frac{\gamma H}{C_u}$.

This notion was later extended with the expression:

$$N = \frac{2\sigma_0}{\sigma_c} \tag{4.18}$$

In the case of an anisotropic initial stress state, the stability number is defined by:

$$N = \frac{3\sigma_0^1 - \sigma_0^3}{\sigma_c} \tag{4.19}$$

where $\sigma_0^1$ and $\sigma_0^3$ are the principal major and minor stresses respectively, in the plane normal to the tunnel axis.

Depending of the value $N$, different situations can be distinguished (Fig. 4.7):

If $N < 1$, no plastic zone develops.
If $1 < N < 2$, A plastic zone appears behind the face.

**Fig. 4.7** Development of the plastic zone in the vicinity of the tunnel face depending on the value of the stability number

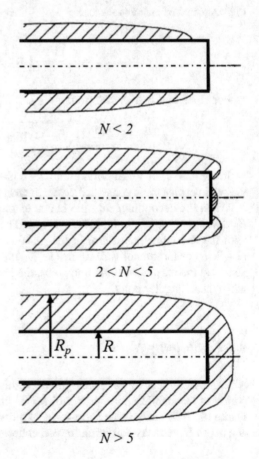

If $2 < N < 5$, The plastic zone starts to develop in the vicinity of the face.
If $N > 5$, The tunnel face is included in the plastic zone which develops in front of the face.

The analysis of the stability conditions of the tunnel face is essential for the choice of the excavation method and the type of support. If $N > 5$, the stability of the face becomes critical, pre-convergence and extrusion become high. It is then necessary to use techniques of pre-support and reinforcement of the face.

For tunnels excavated in low permeability soil below the water table, the analysis of the stability of the face is carried out under short-term conditions.

It should be noted that the notion of stability number defined above applies to sufficiently deep tunnels excavated in a homogeneous medium.

If the face is formed or is located near geological facies having different geome-chanical behaviors, a more in-depth analysis is necessary as the stability conditions can then be more severe than indicated by the value of $N$. As previously mentioned, the proximity of permeable zones under high hydraulic head constitutes a high risk of instability of the tunnel face.

For tunnels located at shallow depths, face instability can trigger a cave-in which reaches the surface and can then lead to major disturbances, especially in urban sites (Fig. 4.8):

The tunnel face stability at shallow depth has given rise to numerous studies both for purely cohesive soils and for frictional soils. Different methods were used as for example:

- limit equilibrium models (Leca and Dormieux 1990; Leca and Panet 1990; Subrin and Wong 2002),
- centrifuge studies (Mair 1979; Schofield 1980; Chambon 1990).
- numerical models (Guilloux et al. 2005; Subrin 2002).

## 4.3 Stress and Displacement Fields for an Elastic Perfectly Plastic Behavior of the Ground

In this section, analytical solutions for the stress and strain fields around a tunnel of circular section excavated in an isotropic elastoplastic medium with an isotropic initial stress state are proposed for various usual plasticity criteria and for a Mohr–Coulomb plastic potential (Panet 1993). These solutions for the displacement fields correspond to the face mode for the plastic flow law, that is to say when the principal stress $\sigma_x$ in the direction of the tunnel axis remains intermediate ($\sigma_r < \sigma_x < \sigma_\theta$). The reader will find in appendix the solution in edge mode ($\sigma_r < \sigma_x = \sigma_\theta$) for the case of a Mohr–Coulomb yield surface.

The stress field in the plastic zone ($R \le r \le R_p$) is obtained from:

- the equilibrium equation in axisymmetry conditions:

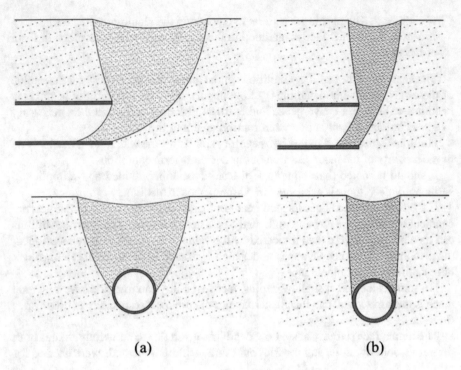

**Fig. 4.8** Instability of the face of a tunnel excavalted at shallow depth with occurrence of a cave-in: **a** clayey ground (after Mair 1979), **b** sandy ground (after Chambon 1990)

$$\frac{d\sigma_r}{dr} + \frac{\sigma_r - \sigma_\theta}{r} = 0 \tag{4.20}$$

- the yield criterium which is fulfilled in every point of the plastic zone:

$$F(\sigma_\theta, \sigma_r) = 0 \tag{4.21}$$

- the boundary conditions:

$$\begin{aligned} &\text{For } r = R, \quad \sigma_r|_{r=R} = (1 - \lambda)\sigma_0 \\ &\text{For } r = R_p, \quad \sigma_r|_{r=R_p} = (1 - \lambda_e)\sigma_0 \end{aligned} \tag{4.22}$$

The strain field in the plastic zone ($R \leq r \leq R_p$) is obtained from:

- the decomposition of the strain rate in an elastic and a plastic part;

$$\dot{\varepsilon} = \dot{\varepsilon}_e + \dot{\varepsilon}_p \tag{4.23}$$

- the plastic flow rule:

$$K_\psi \dot{\varepsilon}_\theta^p + \dot{\varepsilon}_r^p = 0 \tag{4.24}$$

where the dilatancy parameter $K_\psi$ is defined from the dilatancy angle of the material $\psi$:

$$K_\psi = \frac{1 + \sin\psi}{1 - \sin\psi} \tag{4.25}$$

- the compatibility condition for strains:

$$\frac{d\varepsilon_\theta}{dr} = \frac{\varepsilon_r - \varepsilon_\theta}{r} \tag{4.26}$$

In the elastic zone ($r \geq R_p$), the strain and stress fields are given by:

$$\sigma_r = (1 - \lambda_e \frac{R_p^2}{r^2})\sigma_0$$

$$\sigma_\theta = (1 + \lambda_e \frac{R_p^2}{r^2})\sigma_0$$

$$u = \lambda_e \frac{\sigma_0}{2G} \frac{R_p^2}{r} \tag{4.27}$$

## 4.3.1 Tresca Yield Criterion

For Tresca yield criterion, we obtain:

For $R \leq r \leq R_p$ (plastic zone)

$$\frac{\sigma_r}{\sigma_0} = 1 - \lambda + 2\lambda_e \ln\frac{r}{R}$$

$$\frac{\sigma_\theta}{\sigma_0} = 1 - \lambda + 2\lambda_e\left(1 + \ln\frac{r}{R}\right)$$

$$\frac{R_p}{R} = \exp\left(\frac{\lambda - \lambda_e}{2\lambda_e}\right) \tag{4.28}$$

Assuming isochoric deformation ($\nu = 0.5$ and $\psi = 0$), we get:

$$u = \lambda_e \frac{\sigma_0}{2G} \frac{R_p^2}{r} = \frac{C}{2G} \frac{R_p^2}{r}$$

at the wall: $\dfrac{u_R}{R} = \lambda_e \dfrac{\sigma_0}{2G}\left(\dfrac{R_p}{R}\right)^2 = \dfrac{\sigma_0}{2G}\dfrac{1}{N}\exp(\lambda N - 1) \tag{4.29}$

### 4.3.2   Mohr–Coulomb Yield Criterion

Assuming that in the plastic zone, the stresses fulfill the condition of face mode which is $\sigma_r < \sigma_x < \sigma_\theta$ integration of the equilibrium Eq. (4.20) gives:

For $R \leq r \leq R_p$ (plastic zone)

$$\sigma_r = \frac{\sigma_c}{K_p - 1}\left[\left(\frac{r}{R}\right)^{K_p-1} - 1\right] + (1 - \lambda)\sigma_0\left(\frac{r}{R}\right)^{K_p-1}$$

$$\sigma_\theta = \frac{\sigma_c}{K_p - 1}\left[K_p\left(\frac{r}{R}\right)^{K_p-1} - 1\right] + K_p(1 - \lambda)\sigma_0\left(\frac{r}{R}\right)^{K_p-1}$$

$$\frac{R_p}{R} = \left[\frac{2}{K_p + 1}\frac{(K_p - 1)\sigma_0 + \sigma_c}{(1 - \lambda)(K_p - 1)\sigma_0 + \sigma_c}\right]^{\frac{1}{K_p-1}} \tag{4.30}$$

or:

For $R \leq r \leq R_p$ (plastic zone)

$$\sigma_r = \frac{\sigma_0}{K_p - 1}\left[2\lambda_e\left(\frac{r}{R_p}\right)^{K_p-1} - \frac{\sigma_c}{\sigma_0}\right]$$

$$\sigma_\theta = K_p\sigma_r + \sigma_c$$

$$\frac{R_p}{R} = \left[\frac{2\lambda_e}{(K_p + 1)\lambda_e - (K_p - 1)\lambda}\right]^{\frac{1}{K_p-1}} \tag{4.31}$$

A simple approximate solution for the displacement field can be obtained under the assumption that in the plastic zone the elastic part of the strains increment can be neglected:

$$\varepsilon_r = -\frac{\lambda_e\sigma^0}{2G} + \Delta\varepsilon_r \approx -\frac{\lambda_e\sigma^0}{2G} + \Delta\varepsilon_r^p$$

$$\varepsilon_\theta = \frac{\lambda_e\sigma^0}{2G} + \Delta\varepsilon_\theta \approx -\frac{\lambda_e\sigma^0}{2G} + \Delta\varepsilon_\theta^p \tag{4.32}$$

This assumption leads to the following solution:

for $\;\; R \leq r \leq R_p$

$$\frac{u}{r} = \frac{\lambda_e\sigma^0}{2(K_\psi + 1)G}\left[2\left(\frac{R_p}{r}\right)^{K_\psi+1} + K_\psi - 1\right]$$

at the wall: $\;\; \dfrac{u_R}{R} = \dfrac{\lambda_e\sigma^0}{2(K_\psi + 1)G}\left[2\left(\dfrac{R_p}{R}\right)^{K_\psi+1} + K_\psi - 1\right] \tag{4.33}$

However, this simplifying assumption is not necessary and the exact closed form solution can be obtained (see Appendix):

$$\frac{u}{r} = \lambda_e \frac{\sigma_0}{2G}\left[F_1 + F_2\left(\frac{r}{R_p}\right)^{K_p-1} + F_3\left(\frac{R_p}{r}\right)^{K_\psi+1}\right]$$

where:

$$F_1 = -(1-2v)\frac{K_p+1}{K_p-1}$$

$$F_2 = 2\frac{1 + K_p K_\psi - v(K_p+1)(K_\psi+1)}{(K_p-1)(K_p+K_\psi)}$$

$$F_3 = 2(1-v)\frac{K_p+1}{K_p+K_\psi}$$

at the wall:

$$\frac{u_R}{R} = \lambda_e \frac{\sigma_0}{2G}\left[F_1 + F_2\left(\frac{R}{R_p}\right)^{K_p-1} + F_3\left(\frac{R_p}{R}\right)^{K_\psi+1}\right] \tag{4.34}$$

Stress and strain paths at the tunnel wall are shown in Fig. 4.9.

As for example, the radial and hoop stress distribution as a function of the distance to the wall is shown in Fig. 4.10 assuming a friction angle $\phi = 30°$ and a stability number $N = 4$.

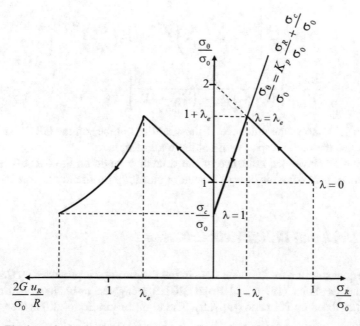

**Fig. 4.9** Elastic-perfectly plastic behavior—Mohr–Coulomb yield criterion. Strain and stress paths at the tunnel wall

**Fig. 4.10** Elastic-perfectly plastic ground with Mohr–Coulomb criterion. Radial and hoop stress for $\lambda = \lambda_e$ and $\lambda = 1(N = 4, \phi = 30°)$

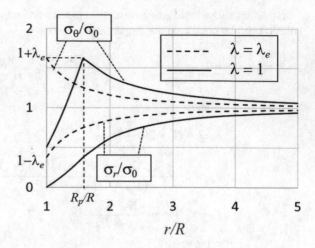

The Ground Reaction Curve (GRC) represents the relation between the radial displacement $u_R$ and the radial stress at the tunnel wall $\sigma_R = (1 - \lambda)\sigma_0$.

For $\lambda \leq \lambda_e$, this curve is a straight line of equation $\sigma_R = \sigma_0 - 2G\frac{u_R}{R}$. For $\lambda > \lambda_e$, depending on the plasticity model that governs the behavior of the ground, we have seen in the previous section that the displacement at the wall can be written as a function of the plastic radius $R_p$ and that this plastic radius is a function of the radial stress at the wall $\sigma_R$. For example for Mohr–Coulomb plasticity model [Eqs. (4.30) and (4.34)]:

$$\frac{u}{R} = \frac{\lambda_e \sigma_0}{2G}\left[F_1 + F_2\left(\frac{R}{R_p}\right)^{K_p-1} + F_3\left(\frac{R_p}{R}\right)^{K_\psi+1}\right]$$

$$\frac{R_p}{R} = \left[\frac{2\lambda_e}{(K_p + 1)\lambda_e - (K_p - 1)\lambda}\right]^{\frac{1}{K_p-1}} \tag{4.35}$$

Figure 4.11 shows the influence of the stability number on the GRC. For higher values for $N$, the radial displacement at the wall is higher.

Figure 4.12 shows the influence of the dilatancy angle on the GRC. The radial displacement is maximal in case of associate plasticity (normality flow rule).

### 4.3.3 Hoek and Brown Yield Criterion

An analytical solution for the stress field in the plastic zone is proposed by Carranza-Torres and Fairhurst (1999) and Rojat (2010) using the normalized form of the stresses proposed by P. Londe (Eq. 4.9). The equilibrium Eq. (4.20) is then written as:

**Fig. 4.11** Ground reaction curve for a tunnel excavated in an elastic perfectly plastic ground with Mohr–Coulomb yield criterion and assuming zero volumetric strain ($\phi = 30°$, $\psi = 0$)- Influence of the stability number $N$

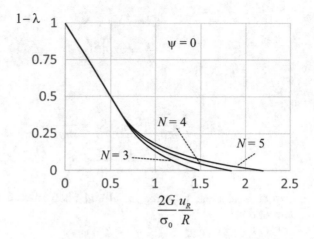

**Fig. 4.12** Ground reaction curve for a tunnel excavated in an elastic perfectly plastic ground with Mohr–Coulomb yield criterion ($\phi = 30°$, $N = 4$)- Influence of the dilatancy angle

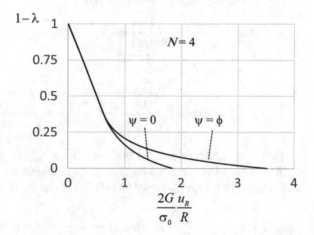

$$\frac{dS_r}{dr} + \frac{S_r - S_\theta}{r} = 0 \qquad (4.36)$$

By introducing the normalized form of the Hoek and Brown criterion (Eq. (4.9), into the equilibrium Eq. (4.36), we obtain the following differential equation:

$$\frac{dS_r}{dr} - \frac{\sqrt{S_r}}{r} = 0 \qquad (4.37)$$

The following expressions are obtained for the stresses and the plastic radius:

For $R \leq r \leq R_p$ (plastic zone)

$$S_r = \left(\sqrt{S_i} + \frac{1}{2}\ln\left(\frac{r}{R}\right)\right)^2$$

$$S_\theta = S_r + \sqrt{S_r}$$

$$\frac{R_p}{R} = \exp\left[2\left(\sqrt{S_r^*} - \sqrt{S_i}\right)\right] \tag{4.38}$$

with:

$$S_r = \frac{\sigma_r}{m\sigma_c} + \frac{s}{m^2}; \quad S_\theta = \frac{\sigma_\theta}{m\sigma_c} + \frac{s}{m^2}$$

$$S_i = \frac{(1-\lambda)\sigma_0}{m\sigma_c} + \frac{s}{m^2}; \quad S_r^* = \frac{(1-\lambda_e)\sigma_0}{m\sigma_c} + \frac{s}{m^2} \tag{4.39}$$

In an equivalent way, the radial and hoop stresses and the plastic radius are expressed as:

For $R \leq r \leq R_p$ (plastic zone)

$$\sigma_r = -\frac{s\sigma_c}{m} + \frac{1}{4m\sigma_c}\left\{m\sigma_c\ln\frac{r}{R} + 2\sqrt{m\sigma_c(1-\lambda)\sigma_0 + s\sigma_c^2 +}\right\}^2$$

$$\sigma_\theta = \sigma_r + \sigma_c\sqrt{m\frac{\sigma_r}{\sigma_c} + s}$$

$$\frac{R_p}{R} = \exp\left[2\left(\sqrt{\frac{(1-\lambda_e)\sigma_0}{m\sigma_c} + \frac{s}{m^2}} - \sqrt{\frac{(1-\lambda)\sigma_0}{m\sigma_c} + \frac{s}{m^2}}\right)\right] \tag{4.40}$$

Under the assumption of a Mohr–Coulomb plastic potential, the displacement field is obtained in the following form for which details of calculations can be found in the paper of Carranza-Torres (2004) and in the Ph.D. thesis of Rojat (2010):

$$u_r(\rho) = \frac{1}{1-A_1}\left(\rho^{A_1} - A_1\rho\right)u_r(1) + \frac{1}{1-A_1}\left(\rho - \rho^{A_1}\right)u_r'(1)$$

$$+ \frac{R_p}{2\tilde{G}}\frac{1}{4}\frac{A_2 - A_3}{1-A_1}\rho(\ln\rho)^2$$

$$+ \frac{R_p}{2\tilde{G}}\left[\frac{A_2 - A_3}{(1-A_1)^2}\sqrt{S_r^*} - \frac{1}{2}\frac{A_2 - A_1A_3}{(1-A_1)^3}\right]\left[\rho^{A_1} - \rho + (1-A_1)\rho\ln\rho\right]$$

where:

$$\rho = \frac{r}{R_p}$$

$$\tilde{G} = \frac{G}{m\sigma_c}$$

$$A_1 = -K_\psi$$

$$A_2 = 1 - v - vK_\psi$$

$$A_3 = v - (1-v)K_\psi$$

$$u_r(1) = \frac{\lambda_e \sigma_0 R_p}{2G}$$

$$u_r'(1) = [(2A_1 - 1) + 2v(1-A_1)]u_r(1) - \frac{R_p}{2\tilde{G}}(A_1 + v(1-A_1))\sqrt{S_r^*} = -u_r(1)$$

(4.41)

at the wall:

$$u_R = \frac{\lambda_e \sigma_0 R_p}{2G} \frac{1}{1-A_1}\left(2\left(\frac{R}{R_p}\right)^{A_1} - (A_1+1)\left(\frac{R}{R_p}\right)\right)$$
$$+ \frac{R_p}{2\tilde{G}} \frac{1}{4} \frac{A_2 - A_3}{1-A_1}\left(\frac{R}{R_p}\right)\left(\ln\left(\frac{R}{R_p}\right)\right)^2$$
$$+ \frac{R_p}{2\tilde{G}}\left[\frac{A_2 - A_3}{(1-A_1)^2}\sqrt{S_r^*} - \frac{1}{2}\frac{A_2 - A_1 A_3}{(1-A_1)^3}\right]$$
$$\left[\left(\frac{R}{R_p}\right)^{A_1} - \left(\frac{R}{R_p}\right) + (1-A_1)\left(\frac{R}{R_p}\right)\ln\left(\frac{R}{R_p}\right)\right]$$

(4.42)

As for example, Ground Reaction Curves are shown in Fig. 4.13 assuming a Hoek & Brown yield criterion, and either plastic incompressibility ($K_\psi = 1$) or Mohr–Coulomb flow rule with a dilatancy angle of 30° ($K_\psi = 3$). In these examples, a value of the stability number $N = 4$ has been considered and the parameters of the Hoek & Brown criterion are: $m = 7$ et $s = 1$.

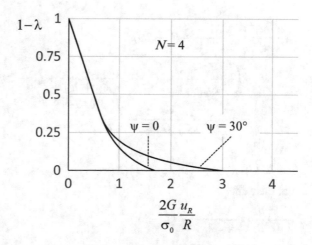

**Fig. 4.13** Ground Reaction Curve for a tunnel excavated in an elastic perfectly plastic ground with Hoek & Brown yield criterion ($N = 4$, $m = 7$, $s = 1$)— Influence of the dilatancy angle

## 4.4  Brittle Failure

For a medium which exhibits brittle failure, the resistance sharply drops beyond the peak strength. This behavior is modelled by considering a discontinuity of the deformation law between the maximum resistance (peak strength) and the residual strength (Panet 1976; Nguyen and Bérest 1979).

It is assumed that the maximum resistance is described by a Mohr–Coulomb criterion:

$$\sigma_1 = K_p \sigma_3 + \sigma_c \tag{4.43}$$

and that the residual strength of the cohesionless yielded rock is given by:

$$\sigma_1 = K_R \sigma_3 \quad \text{with } K_R \leq K_p \tag{4.44}$$

By following the same approach as in the previous section, the stress and displacement fields in the failed zone ($R \leq r \leq R_p$) where the criterion of residual strength is fulfilled are obtained as:

$$\sigma_r = (1 - \lambda)\sigma_0 \left(\frac{r}{R}\right)^{K_R - 1}$$

$$\sigma_\theta = K_R(1 - \lambda)\sigma_0 \left(\frac{r}{R}\right)^{K_R - 1}$$

$$\text{with: } \frac{R_p}{R} = \left(\frac{1 - \lambda_e}{1 - \lambda}\right)^{\frac{1}{K_R - 1}} \tag{4.45}$$

$$\frac{2G}{\sigma_0}\frac{u}{r} = \frac{2\lambda_e}{K_\psi + 1} - (1 - 2\nu)$$

$$+ \left(\lambda_e \frac{K_\psi - 1}{K_\psi + 1} - (1 - \lambda_e)\frac{A}{K_R + K_\psi} + (1 - 2\nu)(K_\psi + 1)\right)\left(\frac{R_p}{r}\right)^{K_\psi + 1}$$

$$+ (1 - \lambda_e)\frac{A}{K_R + K_\psi} - \left(\frac{r}{R_p}\right)^{K_R - 1}$$

with:

$$A = 1 + K_r K_\psi - \nu(K_\psi + 1)(K_R + 1)$$

at the wall:

$$\frac{2G}{\sigma_0}\frac{u}{R} = \frac{2\lambda_e}{K_\psi + 1} - (1 - 2\nu)$$

$$+ \left(\lambda_e \frac{K_\psi - 1}{K_\psi + 1} - (1 - \lambda_e)\frac{A}{K_R + K_\psi} + (1 - 2\nu)(K_\psi + 1)\right)\left(\frac{R_p}{R}\right)^{K_\psi + 1}$$

$$+ (1 - \lambda_e) \frac{A}{K_R + K_\psi} \left( \frac{R}{R_p} \right)^{K_R - 1} \tag{4.46}$$

Note that when the deconfinement ratio $\lambda$ tends to 1, the plastic radius tends to infinity so that there is no possibility of stability without support.

## 4.5 Effect of Gravity on the Stability of the Vault of the Tunnel

In the above analytical solutions for the Ground Reaction Curve, the supporting pressure monotonously decreases with the convergence of the tunnel's wall. This conclusion is however not in agreement with field evidence and practitioner engineers know from experience that a light support quickly installed behind the face can be effective in controlling the convergence, whereas if the ground is left to deform freely with the development of a large yielded zone, a much heavier support will be necessary with the development of much higher support pressures. This apparent contradiction comes from the fact that in the theoretical calculations, for a deep tunnel, gravity was not introduced. Gravity forces are implicitly taken into account in the evaluation of the initial stress state. It may be necessary to examine the vault stability of the decompressed zone (yielded zone) under the effect of gravity.

Pacher (1964) suggested to include an additional vertical load given by the dead weight of the loosened rock in the yielded zone in the determination of the Ground Reaction Curve (Fig. 4.14):

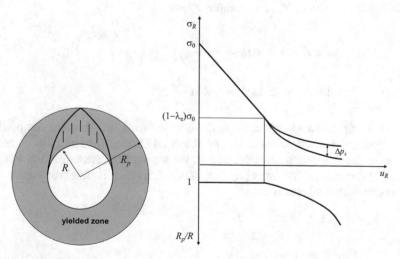

**Fig. 4.14** Additional supporting pressure due to the weight of the yielded zone

$$\Delta p_s = \gamma \left( R_p - R \right) \tag{4.47}$$

where $\gamma$ is the unit weight of the rock mass in the yielded zone.

This correction is however too conservative and can lead to over-dimensioning of the support structure. It can be improved by following the approach of Caquot and Kérisel (1966) (see also Detournay 1984; Panet and Bernardet 1989) in the analysis of the limit equilibrium of the yielded rock mass in the vault of the tunnel assuming that the ground is at residual strength. It consists in solving the equilibrium equation with gravity forces:

$$\frac{d\sigma_r}{dr} + \frac{\sigma_r - \sigma_\theta}{r} + \gamma = 0 \tag{4.48}$$

and zero radial stress at the interface between the plastic zone and the elastic zone as the boundary condition:

$$\text{For } r = R_p \quad \sigma_r|_{r=R_p} = 0 \tag{4.49}$$

In the yielded zone, it is reasonable to consider that the ground's behavior is characterized by the residual strength. For Mohr–Coulomb criterion:

$$\sigma_\theta = K_R \sigma_r \tag{4.50}$$

The integration of the differential Eq. (4.48), with the condition (4.50) and the limit condition (4.49) permits to evaluate the additional radial stress at the vault of tunnel due to the weight of the decompressed zone:

$$\text{for } K_p \neq 2, \ \Delta p_s = \frac{\gamma R}{K_p - 2} \left( 1 - \left( \frac{R}{R_p} \right)^{K_p - 2} \right)$$

$$\text{for } K_p = 2, \ \Delta p_s = \gamma R \ln \frac{R_p}{R} \tag{4.51}$$

When a large yielded zone develops, this correction may dominant as compared to the supporting pressure determined from the usual Ground Reaction Curve. Moreover, note that considering this correction, the supporting pressure at the vault of the tunnel depends on the tunnel radius; thus a size effect is observed which is in accordance with field observations.

## Appendix: Tunnel with Circular Section Excavated in an Elastic Perfectly Plastic Medium with Mohr–Coulomb Yield Criterion. Axi-symmetric Case—Face Mode and Edge Mode

In the plastic zone, the radial displacement and the stress and strain fields can be determined from the following governing equations:

(a) Equilibrium equation:

$$\frac{d\sigma_r}{dr} + \frac{\sigma_r - \sigma_\theta}{r} = 0 \qquad (4.52)$$

(b) Elastoplastic constitutive equations in plane strain:

$$\varepsilon_r = \frac{du}{dr} = \varepsilon_r^e + \varepsilon_r^p = \frac{1}{E}(\sigma_r - \nu\sigma_\theta - \nu\sigma_x) + \varepsilon_r^p$$

$$\varepsilon_\theta = \frac{u}{r} = \varepsilon_r^e + \varepsilon_r^p = \frac{1}{E}(-\nu\sigma_r + \sigma_\theta - \nu\sigma_x) + \varepsilon_\theta^p$$

$$\varepsilon_x = 0 = \varepsilon_r^e + \varepsilon_r^p = \frac{1}{E}(-\nu\sigma_r - \nu\sigma_\theta + \sigma_x) + \varepsilon_x^p \qquad (4.53)$$

(c) Mohr–Coulomb yield criterion:

$$\sigma_\theta - K_p\sigma_r - \sigma_c = 0 \qquad (4.54)$$

(d) Plastic flow rule:

- Face mode ($\sigma_r < \sigma_x < \sigma_\theta$):

$$Q = \sigma_\theta - K_\psi\sigma_r \qquad (4.55)$$

- Edge mode ($\sigma_r < \sigma_x = \sigma_\theta$):

$$Q_1 = \sigma_\theta - K_\psi\sigma_r$$
$$Q_2 = \sigma_x - K_\psi\sigma_r \qquad (4.56)$$

The integration of the equilibrium Eq. (4.52) considering the Mohr–Coulomb criterion (4.54) and the following boundary condition:

$$\text{For } r = R \quad \sigma_r|_{r=R} = (1 - \lambda)\sigma_0 \qquad (4.57)$$

gives the expression for the stresses in the plastic zone:

For $R \leq r \leq R_p$ (plastic zone)

$$\sigma_r = \frac{\sigma_c}{K_p - 1}\left[\left(\frac{r}{R}\right)^{K_p-1} - 1\right] + (1 - \lambda)\sigma_0\left(\frac{r}{R}\right)^{K_p-1}$$

or:

$$\sigma_r = \frac{1}{K_p - 1}\left[2\lambda_e\sigma_0\left(\frac{r}{R_p}\right)^{K_p-1} - \sigma_c\right]$$

$$\sigma_\theta = K_p\sigma_r + \sigma_c \tag{4.58}$$

The plastic radius is given by the following expression:

$$\frac{R_p}{R} = \left[\frac{2\lambda_e}{\left(K_p + 1\right)\lambda_e - \left(K_p - 1\right)\lambda}\right]^{\frac{1}{K_p-1}} \tag{4.59}$$

*Face mode*:
In face mode, the plastic strain rate is given by the plastic flow rule:

$$\dot{\varepsilon}_r^p = \dot{\mu}\frac{\partial Q}{\partial \sigma_r} = -\dot{\mu}K_\psi$$

$$\dot{\varepsilon}_\theta^p = \dot{\mu}\frac{\partial Q}{\partial \sigma_\theta} = \dot{\mu}$$

$$\dot{\varepsilon}_x^p = \dot{\mu}\frac{\partial Q}{\partial \sigma_x} = 0 \tag{4.60}$$

where $\dot{\mu}$ is the plastic multiplier.

From Eqs. (4.53) and (4.60), the following relationships are deduced:

$$\dot{\sigma}_x = v(\dot{\sigma}_r + \dot{\sigma}_\theta)$$

$$\frac{E}{1 - v^2}\frac{d\dot{u}}{dr} = \dot{\sigma}_r - \frac{v}{1 - v}\dot{\sigma}_\theta - \frac{E}{1 - v^2}\dot{\mu}K_\psi$$

$$\frac{E}{1 - v^2}\frac{\dot{u}}{r} = -\frac{v}{1 - v}\dot{\sigma}_r + \dot{\sigma}_\theta + \frac{E}{1 - v^2}\dot{\mu} \tag{4.61}$$

By eliminating the plastic multiplier in the above equations, the following differential equation is obtained:

$$\frac{E}{1 - v^2}\left(\frac{d\dot{u}}{dr} + K_\psi\frac{u}{r}\right) = \left(1 - \frac{v}{1 - v}K_\psi\right)\dot{\sigma}_r + \left(K_\psi - \frac{v}{1 - v}\right)\dot{\sigma}_\theta \tag{4.62}$$

or introducing the elastic shear modulus:

$$2G\left(\frac{d\dot{u}}{dr} + K_\psi\frac{\dot{u}}{r}\right) = \left(1 - v - vK_\psi\right)\dot{\sigma}_r + \left(-v + (1 - v)K_\psi\right)\dot{\sigma}_\theta \qquad (4.63)$$

This equation can be integrated following the loading path corresponding to the progressive stress release of the rock mass from its initial state where in situ stresses are uniform, isotropic and equal to $\sigma_0$ up to the current state $(\sigma_r, \sigma_\theta)$. This permits to rewrite Eq. (4.63) as:

$$2G\left(\frac{du}{dr} + K_\psi\frac{u}{r}\right) = \left(1 - v - vK_\psi\right)\Delta\sigma_r + \left(-v + (1 - v)K_\psi\right)\Delta\sigma_\theta \qquad (4.64)$$

where:

$$\Delta\sigma_r = \sigma_r - \sigma_0$$
$$\Delta\sigma_\theta = \sigma_\theta - \sigma_0 = K_p\Delta\sigma_r + (K_p + 1)\lambda_e\sigma_0 \qquad (4.65)$$

By introducing the relationships (4.65) into the differential Eq. (4.64), we get:

$$\frac{2G}{r^{K_\psi}}\frac{d}{dr}\left(r^{K_\psi}u\right) = \left(1 + K_\psi K_p - v(K_p + 1)(K_\psi + 1)\right)(\sigma_r - \sigma_0)$$
$$+ \left[K_\psi - v(K_\psi + 1)\right](K_p + 1)\lambda_e\sigma_0 \qquad (4.66)$$

The above differential equation can be integrated by replacing $\sigma_r$ by its expression (Eq. 4.58) with the following boundary condition:

$$\text{For } r = R_p, \ u = \lambda_e\frac{\sigma_0}{2G}R_p \qquad (4.67)$$

We get:

$$\frac{u}{r} = \lambda_e\frac{\sigma_0}{2G}\left[F_1 + F_2\left(\frac{r}{R_p}\right)^{K_p-1} + F_3\left(\frac{R_p}{r}\right)^{K_\psi+1}\right]$$

where:

$$F_1 = -(1 - 2v)\frac{K_p + 1}{K_p - 1}$$

$$F_2 = 2\frac{1 + K_pK_\psi - v(K_p + 1)(K_\psi + 1)}{(K_p - 1)(K_p + K_\psi)}$$

$$F_3 = 2(1 - v)\frac{K_p + 1}{K_p + K_\psi} \qquad (4.68)$$

One can easily check that $F_1 + F_2 + F_3 = 0$ as for $r = R_p$, $\frac{u}{R_p} = \lambda_e\frac{\sigma_0}{2G}$.

*Edge mode:*

During the development of the plastic zone in face mode, it may happen that at the tunnel wall, the medium enters in edge mode for $\lambda = \lambda_a > \lambda_e$, such that:

$$\sigma_R < \sigma_\theta = \sigma_x \tag{4.69}$$

for $\lambda_a < \lambda < 1$, a plastic zone of radius $R_a$ develops in edge mode:

$$R < R_a < R_p \tag{4.70}$$

For $\lambda = \lambda_a$ and for $r = R:\Delta\sigma_x = \Delta\sigma_\theta = v(\Delta\sigma_R + \Delta\sigma_\theta)$

with : $\Delta\sigma_R = -\lambda_a\sigma_0$ $\tag{4.71}$

Thus:

$$\Delta\sigma_\theta = -\frac{v}{1-v}\lambda_a\sigma_0 \tag{4.72}$$

and considering Eq. (4.65):

$$\lambda_a = \frac{(1-v)(K_p+1)}{K_p - v(K_p+1)}\lambda_e \tag{4.73}$$

One can easily check that $\lambda_a > \lambda_e$ and the condition for the development of a plastic zone in edge mode $\lambda_a < 1$ can be written as:

$$\lambda_e < \frac{K_p - v(K_p+1)}{(1-v)(K_p+1)} \tag{4.74}$$

Table 4.1 gives the maximum value of $\lambda_e$ so that $\lambda_a < 1$, as a function of the Poisson's ratio $v$ and of $K_p$:

In edge mode, the plastic flow rule is applied as a linear combination of the flow rules corresponding to two adjacent faces. Plastic strain rates are thus given by:

**Table 4.1** Maximal value of $\lambda_e$ so that $\lambda_a < 1$

| $\phi$ (°) | 0 | 20 | 30 | 35 | 40 | 45 | 50 |
|---|---|---|---|---|---|---|---|
| $K_p$ | 1 | 2.04 | 3.00 | 3.69 | 4.60 | 5.83 | 7.55 |
| $v = 0$ | 0.50 | 0.67 | 0.75 | 0.79 | 0.82 | 0.85 | 0.88 |
| $v = 0.2$ | 0.38 | 0.59 | 0.69 | 0.73 | 0.78 | 0.82 | 0.85 |
| $v = 0.3$ | 0.29 | 0.53 | 0.64 | 0.70 | 0.74 | 0.79 | 0.83 |
| $v = 0.5$ | 0 | 0.34 | 0.50 | 0.57 | 0.64 | 0.71 | 0.77 |

$$\dot{\varepsilon}_r^p = \dot{\mu}_1 \frac{\partial Q_1}{\partial \sigma_r} + \dot{\mu}_2 \frac{\partial Q_2}{\partial \sigma_r} = -(\dot{\mu}_1 + \dot{\mu}_2) K_\psi$$

$$\dot{\varepsilon}_\theta^p = \dot{\mu}_1 \frac{\partial Q_1}{\partial \sigma_\theta} + \dot{\mu}_2 \frac{\partial Q_2}{\partial \sigma_\theta} = \dot{\mu}_1 \qquad (4.75)$$

$$\dot{\varepsilon}_x^p = \dot{\mu}_1 \frac{\partial Q_1}{\partial \sigma_x} + \dot{\mu}_2 \frac{\partial Q_2}{\partial \sigma_x} = \dot{\mu}_2$$

From Eqs. (4.53) and (4.75), the following relationships can be deduced:

$$E \frac{d\dot{u}}{dr} = \dot{\sigma}_r - \nu \dot{\sigma}_\theta - \nu \dot{\sigma}_x - E(\dot{\mu}_1 + \dot{\mu}_2) K_\psi$$

$$E \frac{\dot{u}}{r} = -\nu \dot{\sigma}_r + \dot{\sigma}_\theta - \nu \dot{\sigma}_x + E \dot{\mu}_1 \qquad (4.76)$$

$$0 = -\nu \dot{\sigma}_r - \nu \dot{\sigma}_\theta + \dot{\sigma}_x + E \dot{\mu}_2$$

and by eliminating the plastic multipliers $\dot{\mu}_1$ and $\dot{\mu}_2$:

$$E\left(\frac{d\dot{u}}{dr} + \frac{\dot{u}}{r}\right) = \left(1 - 2\nu K_\psi\right)\dot{\sigma}_r + \left(-\nu + K_\psi(1 - \nu)\right)(\dot{\sigma}_\theta + \dot{\sigma}_x) \qquad (4.77)$$

Like for the face mode, this equation can be integrated following the loading path corresponding to the progressive stress release of the rock mass from its initial state where in situ stresses are uniform, isotropic and equal to $\sigma_0$ up to the current state $(\sigma_r, \sigma_\theta)$. We obtain:

$$\frac{E}{r^{K_\psi}} \frac{d}{dr}\left(r^{K_\psi} u\right) = \left(1 - 2\nu\left(K_\psi + K_p\right) + 2(1 - \nu)K_p K_\psi\right)(\sigma_r - \sigma_0)$$
$$+ 2\left[(1 - \nu)K_\psi - \nu\right]\left(K_p + 1\right)\lambda_e \sigma_0 \qquad (4.78)$$

Integrating this equation with the boundary condition (4.67) gives:

$$\frac{u}{r} = \lambda_e \frac{\sigma_0}{2G}\left[A_1 + A_2\left(\frac{r}{R_p}\right)^{K_p - 1} + A_3\left(\frac{R_p}{r}\right)^{K_\psi + 1}\right]$$

où :

$$A_1 = -\frac{1 - 2\nu}{1 + \nu} \frac{K_p + 1}{K_p - 1} \frac{2K_\psi + 1}{2K_\psi + 1} \qquad (4.79)$$

$$A_2 = 2\frac{1 + 2K_p K_\psi - 2\nu\left(K_p + K_\psi + K_p K_\psi\right)}{(1 + \nu)\left(K_p - 1\right)\left(K_p + K_\psi\right)}$$

$$A_3 = (F_1 - A_1)\left(\frac{R_a}{R_p}\right)^{K_\psi + 1} + (F_2 - A_2)\left(\frac{R_a}{R_p}\right)^{K_p + K_\psi} + F_3$$

where $F_1$, $F_2$, $F_3$ are the constant determined in the face mode (Eq. (4.68) and:

$$\frac{R_a}{R} = \left[ \frac{(1 - 2\nu)(K_p + 1)\lambda_e}{[(1 - \nu)K_p - \nu][(K_p + 1)\lambda_e - (K_p - 1)\lambda]} \right]^{\frac{1}{K_p - 1}} \qquad (4.80)$$

# References

Broms BB, Bennermark H (1967) Stability of clay at vertical openings. J Soil Mech Found Div ASCE 93(1):71–94

Caquot A, Kérisel J (1966) Traité de Mécanique des Sols, 4ème édn. Gauthier-Villars, Paris

Carranza-Torres C (2004) Elasto-plastic solution of tunnel problems using the generalized form of the Hoek-Brown failure criterion. Int J Rock Mech Min Sci 41(3):629–639

Carranza-Torres C, Fairhurst C (1999) The elasto-plastic response of underground excavations in rock masses that satisfy the Hoek-Brown failure criterion. Int J Rock Mech Min Sci 36(6):777–809

Chambon P (1990) Etude sur modèles réduits centrifugés. Application aux tunnels à faible profondeur en terrain meuble pulvérulent. Thèse de doctorat. Université de Nantes

Detournay E (1984) The effect of gravity on the stability of a deep tunnel. Int J Rock Mech Min Sci 21:349–351

Guilloux A, Kazmierczak J-B, Kurdts A, Regal G, Wong H (2005) Stabilité Et Renforcement Des Fronts De Taille Des Tunnels : Une Approche Analytique En Contraintes-Déformation. Tunnels Et Ouvrages Souterrains 188:98–108

Hoek E, Brown ET (1980) Empirical strength criterion for rock masses. J Geotech Eng Div ASCE 106(GT9):1013–1035

Hoek E, Carranza-Torres C, Corkum B (2002) Hoek-brown failure criterion—2002 edition. In: Proceedings of NARMS-TAC 2002, mining innovation andtechnology, Toronto, pp 267–273

Leca E, Dormieux L (1990) Upper and lower bound solutions for the face of shallow circular tunnel in frictional material. Géotechnique 40(4):581–606

Leca E, Panet M (1990) Application du calcul à la rupture à la stabilité du front de taille d'un tunnel. Rev Fr Géotech 43:5–19

Londe P (1988) Discussion on the determination of the shear stress failure in rock masses. ASCE J Geotech Eng Div 114(3):374–376

Mair J (1979) Centrifugal modelling of tunnel construction in clay. Ph.D. Cambridge University

Nguyen Minh D, Bérest P (1979) Etude de la stabilité des cavités souterraines avec un modèle élastoplastique radoucissant. In: Proceedings of 4th international congress on international society rock mechanics, Montreux, vol 1, Balkema, Rotterdam, pp 249–255

Pacher F (1964) Deformationsmessungen im Versuchsstollen als Mittel zur Erforschung des Gebirgsverhaltens und zur Bemessung des Ausbaues. In: Müller L (eds) Grundfragen auf dem Gebiete der Geomechanik/Principles in the field of geomechanics. Felsmechanik und Ingenieurgeologie/Rock mechanics and engineering geology, vol 1. Springer, Berlin, Heidelberg

Panet M (1976) Analyse de la stabilité d'un tunnel creusé dans un massif rocheux en tenant compte du comportement après la rupture. Rock Mech 8:209–223

Panet M, Bernardet A (1989) Les méthodes d'équilibre limite et la méthode convergence-confinement pour le calcul des tunnels peu profonds en sol cohérent. In: Tunnels et micro-tunnels en terrain meuble: du chantier à la théorie. Presses de l'ENPC, pp 645–653

Panet M (1993) Understanding deformations in tunnels. In: Comprehensive rock engineering, vol 1, 27. Pergamon Press, London, pp 663–690

Rojat F (2010) Comportement des tunnels dans les milieux rocheux de faibles caractéristiques mécaniques. Thèse de doctorat, Ecole Nationale des Ponts et Chaussées

Schofield AN (1980) Cambridge geotechnical centrifuge operations. Géotechnique 30(3):227–268

Subrin D, Wong H (2002) Stabilité du front d'un tunnel en milieu frottant: un nouveau mécanisme de rupture 3D. Comptes Rendus à L'académie Des Sciences Mécanique 330:513–519

Subrin D (2002) Etudes théoriques sur la stabilité et le comportement des tunnels renforcés par boulonnage. Thèse de doctorat de l'Ecole Nationale des Travaux Publics

# Chapter 5
# Longitudinal Displacement Profile

The application of the Convergence-Confinement method is based on the knowledge of three curves which are used to calculate the equilibrium state between the ground and the support. These three curves are the *Longitudinal Displacement profile* (LDP) which represents the radial displacement at the tunnel wall as a function of the distance from the face, the *Ground Reaction Curve* (GRC) which is the convergence curve of the unsupported tunnel, and the *Support Confinement Curve* (SCC) (Fig. 5.1).

The determination of the convergence of the ground at the moment when the support becomes active, that is to say when it begins to exert a confining pressure to limit the convergence of the ground, is an essential step of the convergence-confinement method. This convergence is directly related to the unsupported distance (distance of the support installation) $d$, and is obtained from the corresponding radial displacement $u_d$. This displacement $u_d$ corresponds to the deconfinement ratio $\lambda_d$. Knowing the LDP curve makes it possible to optimize the distance and the moment chosen for the installation of the support. Installing the support too early or too near from the face can lead to a significant overload of the support. Conversely, late installation of the support and/or too large unsupported distance can lead to the development of a large failure zone around the tunnel and consequently to deformations that may become impossible to control.

For axi-symmetric conditions, the mathematical expression of the LDP curve is of the following form:

$$u_x = u_R(x, K_{sn}) \tag{5.1}$$

where $u_x$ is the radial displacement at the tunnel wall at the distance $x$ from the face and $K_{sn}$ is the normal stiffness of the support.

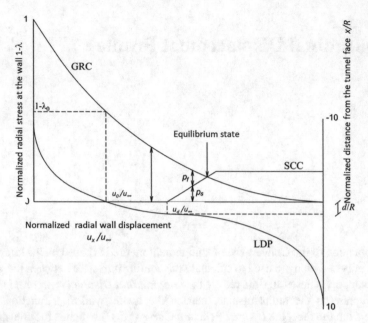

**Fig. 5.1** Schematic representation of the Longitudinal Displacement Profile, the Ground Reaction Curve and the Support Confinement Curve for the application of the Convergence Confinement Method

## 5.1  Longitudinal Displacement Profile for an Unsupported Tunnel

Various empirical expressions of the Longitudinal Displacement Profile have been proposed on the basis of 3D numerical simulations. They generally require knowledge of the displacement at the face and of the final displacement once the excavation is complete.

For an unsupported tunnel excavated in an elastic medium, the mathematical expression of the LDP curve can be written:

$$u_x = a(x)u_{\infty,el}, \quad u_{\infty,el} = \frac{\sigma_0 R}{2G} \tag{5.2}$$

where $u_{\infty,el}$ is the ultimate radial displacement at the wall and $a(x)$ is a dimensionless shape coefficient for which one commonly uses an empirical formula fitted from finite elements numerical simulations (see Sect. 3.1):

$$a(x) = a_0 + (1 - a_0)\left(1 - \left[\frac{mR}{mR + x}\right]^2\right) \tag{5.3}$$

Commonly used values for parameters $a_0$ and $m$ are: $a_0 = 0.25$ and $m = 0.75$.

The deconfinement ratio $\lambda_d$ at the distance of the support installation $d$ is thus given by:

$$\lambda_d = a_0 + (1 - a_0)a_d \text{ with } a_d = \left(1 - \left[\frac{mR}{mR+d}\right]^2\right) \tag{5.4}$$

For an elastoplastic behaviour of the ground, the LDP depends on the extent of the plastic zone. The formulation of the mathematical expression of the LDP can be extended to the elastoplastic case by applying the Self Similarity Principle (SPP) as proposed by Corbetta et al. (1991). It consists in applying a homothetic transformation to the LDP of a tunnel excavated in an elastic ground to get the one for an elastoplastic ground:

$$u_x = \frac{1}{\xi}a(\xi x)u_{\infty,el}, \quad \xi = \frac{u_{\infty,el}}{u_{\infty,ep}} \tag{5.5}$$

where $u_{\infty,ep}$ is the ultimate radial displacement at the wall.

For an unsupported distance $d$, the expression of the shape coefficient as given by Eq. (5.3) leads to:

$$u_d = \frac{1}{\xi}\left(a_0 + (1 - a_0)\left(1 - \left[\frac{mR}{mR+\xi d}\right]^2\right)\right)\frac{\sigma_0 R}{2G} \tag{5.6}$$

Vlachopoulos and Diederichs (2009) have proposed an alternative formulation of the LDP of the following form:

$$u_0^* = \frac{u_0}{u_\infty} = \frac{1}{3}e^{-0,15R_p^*} \quad \text{at the face}$$

$$u_x = u_0 e^{x^*} \text{ for } x^* = \frac{x}{R} \leq 0 \text{ (ahead of the face)}$$

$$u_x = u_\infty\left(1 - (1 - u_0^*)e^{-\frac{3}{2}\frac{x^*}{R_p^*}}\right) \text{ for } x^* = \frac{x}{R} \geq 0 \text{ (behind the face)} \tag{5.7}$$

where $R_p^*$ is the maximum (ultimate) normalized radius of the plastic zone $R_p^* = R_p/R$, ($R$ is the tunnel radius). These authors have obtained these expressions by fitting of results of numerical simulations for an elastic perfectly plastic behavior, assuming zero dilatancy (plastic incompressibility).

In order to compare these different expressions of the LDP curve, we consider the following numerical example:

Tunnel of circular cross section with radius $R = 3.6$ m.

Elastic perfectly plastic ground with zero dilatancy

$$E = 1600 \text{ MPa} ; \nu = 0.3$$
$$\sigma_c = 11.5 \text{ MPa} ; \phi = 26° \quad ; \quad \psi = 0$$

Assuming that the ground remains elastic, the LDP curve obtained from Eq. (5.3) (with $a_0 = 0.25$ and $m = 0.75$) and the one obtained from Eq. (5.7) given by Vlachopoulos and Diederichs (2009) are shown in Fig. 5.2. These curves are compared to the results of an elastic 3D numerical simulation performed with FLAC 3D software. These results show that expression (5.3) gives a very good approximation of the 3D numerical results.

The same comparison is made in the elastoplastic case (stability number $N = 2.8$). Here again, expression (5.3) combined with Corbetta's similarity principle ($a_0 = 0.25$; $m = 0.75$; $\xi = 0.741$) gives the best approximation of the 3D numerical results (Fig. 5.3).

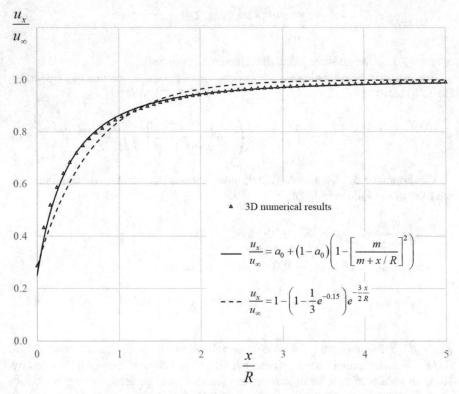

**Fig. 5.2** Tunnel with circular cross section excavated in an elastic ground—Comparison of the Longitudinal Displacement Profiles (LDP) obtained from 3D numerical simulation and from empirical expressions (5.3) ($a_0 = 0.25$ and $m = 0.75$) and (5.7)

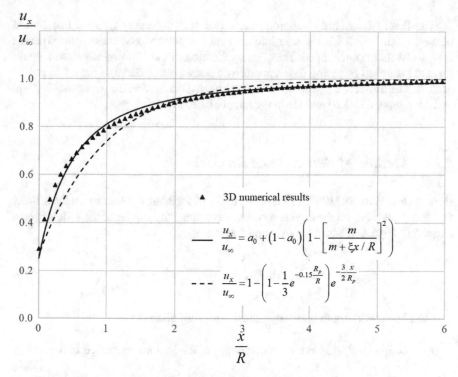

**Fig. 5.3** Tunnel with circular cross section excavated in an elastoplastic ground ($N = 2.8$)—Comparison of the Longitudinal Displacement Profiles (LDP) obtained from 3D numerical simulation and from empirical expressions (5.5) ($a_0 = 0.25; m = 0.75; \xi = 0.741$) and (5.7) ($R_p/R = 1.454$)

## 5.2 Longitudinal Displacement Profile for a Supported Tunnel

The presence of a support affects the Longitudinal Displacement Profile and leads to a lower radial displacement at the wall at the moment when the supported is installed as compared to the one obtained in the unsupported case. Different authors have tried to adjust the empirical formulas proposed above in order to take into account the presence of the support.

The 'implicit' methods permit to obtain the shape coefficient of the LDP of a supported tunnel $a_s(x)$ from that of an unsupported one $a(x)$ by applying an affine transformation of the variable $x$:

$$a_s(x) = a(\alpha x) \tag{5.8}$$

where the factor $\alpha$ depends on the normal stiffness of the lining $K_{sn}$.

Note that in these implicit approaches, only the displacement $u_{sd}$ at the installation of the support is adjusted to take into account the presence of the support whereas the GRC remains unchanged. Thus, the application of the convergence-confinement method is carried out using the characteristic convergence curve of the unsupported ground. The limits of these approaches have been discussed in the papers of Cantieni and Anagnostou (2009) and De la Fuente et al. (2019).

### 5.2.1 Method of Bernaud and Rousset

Bernaud and Rousset (1992, 1996) have proposed various expressions for the affinity factor $\alpha$ depending on the assumed constitutive plasticity model of the ground and a simplified approach based on an average factor $\alpha_{moy}$:

$$\alpha_{moy} = 1.82\sqrt{k_s^*}, \text{ with } k_s^* = \frac{K_{sn}}{E}, \text{ for } k_s^* \geq \frac{R^2}{\left(1.82R_p\right)^2} \tag{5.9}$$

For a Mohr–Coulomb plasticity model with zero dilatancy:

$$\alpha = \alpha_{moy} + 0.035\phi, \phi \text{ is the friction angle of the ground expressed in degrees} \tag{5.10}$$

For a Hoek and Brown plasticity model with zero dilatancy:

$$\alpha = \alpha_{moy} + 0.15m \tag{5.11}$$

For a generalized Hoek and Brown plasticity model with zero dilatancy:

$$\alpha = \alpha_{moy} + 0.25m_b \tag{5.12}$$

These expressions have been established for an incompressible ground and taking $a_0 = 0.27$ and $m = 0.84$. This approach can be applied for values of the stability number $N \leq 5$.

The radial displacement at distance of support installation $u_{ds}$ is thus written as:

$$u_{sd} = u_0 + \frac{a(\xi\alpha d) - a(0)}{1 - a(0)}(u_{s\infty} - u_0) \tag{5.13}$$

This expression contains the ultimate radial displacement of the supported tunnel $u_{s\infty}$, which leads to the resolution of an implicit system of equations when applying the convergence-confinement method. This resolution is thus performed iteratively.

Assuming an elastic behavior of the ground (Sect. 3.3.1), we can develop an explicit solution for $u_{sd}$:

The final equilibrium state is defined by the support pressure $p_s$ and the ultimate displacement $u_{s\infty}$ which are solutions of the following system of equations:

$$p_s = \sigma_0 - 2G\frac{u_{s\infty}}{R}$$

$$p_s = K_s\left(\frac{u_{s\infty}}{R} - \frac{u_{sd}}{R}\right) \tag{5.14}$$

Equations (5.13) and (5.14) give the displacement at the installation of the support $u_{sd}$ and the corresponding deconfinement ratio $\lambda_d$:

$$\frac{u_{sd}}{R} = \frac{b_d + (1 - b_d)(1 + k_{sn})a(0)}{1 + k_{sn}(1 - b_d)}\frac{\sigma_0}{2G}, \lambda_d = \frac{b_d + (1 - b_d)(1 + k_{sn})a(0)}{1 + k_{sn}(1 - b_d)}$$

$$\text{with } k_{sn} = \frac{K_{sn}}{2G}, b_d = \frac{a(\alpha d) - a(0)}{1 - a(0)}$$

$$\tag{5.15}$$

The support pressure at equilibrium is given by:

$$p_s = k_{sn}\frac{(1 - a(0))(1 - b_d)}{1 + (1 - b_d)k_{sn}}\sigma_0 \tag{5.16}$$

Table 5.1 gives the values of the deconfinement ratio $\lambda_d$ for $k_{sn} < 40$ and $d$ with values ranging between $0.25R$ and $2R$. For an elastic ground, this table shows that for $d > 1.5R$, the error is less than 2% when taking the value of $\lambda_d$ corresponding to the unsupported case ($k_{sn} = 0$).

**Table 5.1** Method of Bernaud-Rousset for an elastic ground: Values of the deconfinement ratio $\lambda_d$ as a function of the distance from the face $d$ and of the relative stiffness of the support $k_{sn}$

| $k_{sn}$ | $d/R$ | | | | | | | |
|---|---|---|---|---|---|---|---|---|
| | 0.25 | 0.5 | 0.75 | 1 | 1.25 | 1.5 | 1.75 | 2 |
| 0 | 0.57 | 0.71 | 0.80 | 0.85 | 0.88 | 0.91 | 0.92 | 0.94 |
| 0.25 | 0.55 | 0.70 | 0.78 | 0.84 | 0.87 | 0.90 | 0.92 | 0.93 |
| 0.50 | 0.53 | 0.68 | 0.77 | 0.83 | 0.87 | 0.90 | 0.91 | 0.93 |
| 1 | 0.52 | 0.67 | 0.76 | 0.82 | 0.86 | 0.89 | 0.91 | 0.93 |
| 2 | 0.50 | 0.65 | 0.75 | 0.82 | 0.86 | 0.89 | 0.91 | 0.93 |
| 5 | 0.48 | 0.64 | 0.75 | 0.82 | 0.86 | 0.89 | 0.92 | 0.93 |
| 10 | 0.46 | 0.64 | 0.75 | 0.82 | 0.87 | 0.90 | 0.92 | 0.94 |
| 20 | 0.44 | 0.62 | 0.74 | 0.82 | 0.87 | 0.90 | 0.92 | 0.94 |
| 40 | 0.43 | 0.61 | 0.74 | 0.82 | 0.87 | 0.90 | 0.92 | 0.94 |

### 5.2.2  Method of Nguyen-Minh and Guo

Nguyen-Minh and Guo (1996) have proposed another implicit approach allowing to calculate the radial displacement at the distance of the support installation $d$, for a supported tunnel $u_{ds}$ from its value obtained for an unsupported tunnel $u_d$:

$$\frac{u_{sd}}{u_d} = 0.55 + 0.45\frac{u_{s\infty}}{u_\infty} - 0.42\left(1 - \frac{u_{s\infty}}{u_\infty}\right)^3 \tag{5.17}$$

where $u_\infty$ and $u_{s\infty}$ are the displacements at final equilibrium of the unsupported and supported tunnel respectively.

In the case of a linear elastic behavior of the ground,

$$u_{sd} = \lambda_d \frac{\sigma_0 R}{2G}$$

$$u_d = (a_0 + (1 - a_0)a_d)\frac{\sigma_0 R}{2G} \tag{5.18}$$

the application of the convergence-confinement method gives:

$$u_{s\infty} = \frac{1 + k_{sn}\lambda_d}{1 + k_{sn}}u_\infty \tag{5.19}$$

From Eqs. (5.17), (5.18) and (5.19), we obtain $\lambda_d$ as the solution of the following polynomial equation of degree 3:

$$\frac{1}{a_0 + (1 - a_0)a_d}\lambda_d + \frac{k_{sn}}{1 + k_{sn}}(1 - \lambda_d)\left[0.45 + 0.42\left(\frac{k_{sn}}{1 + k_{sn}}(1 - \lambda_d)\right)^2\right] - 1 = 0 \tag{5.20}$$

Table 5.2 gives the value of $\lambda_d$ as a function of the distance from the face $d$ and of the relative stiffness of the support $k_{sn}$.

### 5.2.3  Comparison of the Accuracy of the Implicit Methods and of the Empirical Expressions

In a recent publication, De La Fuente et al. (2019) have discussed the accuracy of these different methods from comparisons with the results of 3D numerical computations performed with the finite difference code FLAC3D (Itasca 2005). This comparison was made for an elastic perfectly plastic ground and an elastic support. Based on a sensitivity analysis performed on the mechanical properties of the ground and of

**Table 5.2** Method of Nguyen-Minh and Guo for an elastic ground: Values of the deconfinement ratio $\lambda_d$ as a function of the distance from the face $d$ and of the relative stiffness of the support $k_{sn}$

| $k_{sn}$ | $d/R$ | | | | | | |
|---|---|---|---|---|---|---|---|
| | 0.25 | 0.5 | 0.75 | 1 | 1.25 | 1.5 | 2 |
| 0 | 0.58 | 0.73 | 0.81 | 0.86 | 0.89 | 0.91 | 0.94 |
| 0.25 | 0.56 | 0.71 | 0.79 | 0.85 | 0.88 | 0.90 | 0.93 |
| 0.50 | 0.54 | 0.70 | 0.78 | 0.84 | 0.87 | 0.90 | 0.93 |
| 1 | 0.51 | 0.67 | 0.77 | 0.83 | 0.86 | 0.89 | 0.92 |
| 2 | 0.48 | 0.65 | 0.75 | 0.81 | 0.85 | 0.88 | 0.91 |
| 5 | 0.44 | 0.62 | 0.72 | 0.79 | 0.84 | 0.86 | 0.91 |
| 10 | 0.38 | 0.60 | 0.70 | 0.78 | 0.83 | 0.86 | 0.91 |

the support [stability number $N = 2\sigma_0/\sigma_c$, $E^* = E/E_s$ where $E_s$ is the Young's modulus of the lining) it was shown that for a stiff lining installed at about one diameter from the face, as is the case for tunnel boring machine (TBM) excavations, and assuming incompressible plastic deformations, the implicit method of Nguyen-Minh and Guo (Eq. (5.17)] combined with expression (5.3) for the LDP curve gives the best results. These expressions can also be used in the case of a ground with a dilatancy angle $\psi = \phi/3$, as is often assumed in practice. However, for an associated plastic behavior ($\psi = \phi$), the various implicit methods reach their limit.

These authors have also proposed an empirical expression for the displacement at the wall $u_\infty$ and the support pressure $p_{s\infty}$ at equilibrium as a function of the parameters $N$ (with values between 1 and 5), $E^*$ (with values between 0.05 and 1), $\phi$ (friction angle of the ground in degrees), $\psi$ (dilatancy angle of the ground in degrees), and $R^*$ (ratio of the radius of the excavation $R$ to thickness of the lining $e$):

$$\frac{2G}{\sigma_0}\frac{u_\infty}{R} = 1.6244 + 0.012R^* + \phi(1.3 \times 10^{-5}\phi^2 - \frac{0.27}{N})$$

$$+ N\left(0.0178E^* + 0.01855(\psi + 1) + \frac{0.543}{\psi + 1}\right.$$

$$\left. -0.017\phi + \frac{5}{\phi} - \frac{21.99 - 4.076N + 0.24N^2}{\phi(\psi + 1)}\right)$$

$$+ \frac{\psi + 1}{\phi}(-0.0146N^3 + 0.323N^2 - 0.99N)$$

Let $F = 0.922 + 0.0224R^*$

$$+ N\left(\frac{3.88}{\phi} + 9.66 \times 10^{-4)}(\psi + 1) - 0.063\right) + 0.365\frac{E^*}{N} - 0.76\log_{10}\left(100E^*\right)\right)$$

For $F < 0.4$ : $\frac{p_{s\infty}}{\sigma_0} = 0.42 + 0.004\phi + R^*\left(0.0082 - 0.0096\frac{E^*}{N}\right)$

$$- N\left(0.123 + \frac{1}{\phi}\left(0.0685N + \frac{64.57}{\phi} - 7.79\right) - 0.000174(\psi + 1)\right)$$

$$+ E^* \left( \frac{0.0027}{E^{*3}} + \frac{0.1954}{N} + \frac{\psi + 1}{\phi} \left( \frac{-0.1}{N} + 0.0916 \right) \right) - 0.34550 \log_{10}(100E^*)$$

Pour $0.4 < F < 0.8$ : $\dfrac{p_{s\infty}}{\sigma_0} = 1.1149 + 0.0227 R^*$

$$+ \psi(0.0038 - 0.0001\psi) + \frac{0.04}{(\psi + 1)^2}$$

$$- N \left( 0.0879 + \frac{0.00826}{E^*} - 0.000148 \frac{N}{E^{*2}} \right.$$

$$+ 0.158 \frac{N}{\phi} + \frac{41.785}{\phi^2} + \frac{4.06}{E^*\phi} - 0.000463(\psi + 1) - \frac{8.3}{\phi} \right)$$

$$+ E^* \left( -\frac{0.253}{N\phi} + \frac{0.244}{\phi} \right)(\psi + 1) - 0.96 \log_{10}(100E^*)$$

Pour $F \geq 0.8$ : $\dfrac{p_{s\infty}}{\sigma_0} = 0.9617 - 0.0143\phi + 0.0458 R^* - \dfrac{194.85}{\phi^2} + \dfrac{0.0647}{(\psi + 1)^2}$

$$+ N \left( -0.06 \frac{N}{\phi} + \frac{69.55}{\phi^2} - 0.0000357(\psi + 1)^2 \right.$$

$$+ 0.00192(\psi + 1) + \frac{0.095}{E^*\phi} - \frac{1.303}{E^*\phi^2} \right)$$

$$+ E^* \left( -0.202 E^* + \frac{0.000267}{E^{*3}} + 0.478 \frac{\psi + 1}{\phi} \right)) - 0.675 \log_{10}(100E^*) \quad (5.21)$$

These expressions, which are valid for a distance of support installation equal to one diameter of the tunnel cross section, directly give the equilibrium state for a large range of ground parameters and are useful for preliminary design of tunnels excavated with a single-shield TBM.

As for example, we present a comparison of the results obtained by the different methods in terms of the displacement $u_\infty$ and support pressure $p_{s\infty}$ at equilibrium. These simulations are based on the assumptions of the computations presented in the paper De La Fuente et al. (2019) for various relative stiffness of the lining. The ground is assumed to be elastic, perfectly plastic with Mohr–Coulomb criterion. The results for the case of a plastic incompressibility are presented in Table 5.3 and those for the case of associate plasticity in Table 5.4.

Ground properties:

$$N = 2$$
$$\nu = 0.25$$
$$\phi = 30° \quad ; \quad \psi = 0$$

Support properties:

$$\nu_s = 0.2$$
$$\frac{e}{R} = 0.1 \quad ; \quad \frac{d}{2R} = 1$$

**Table 5.3** Comparison of the various implicit and empirical approaches with reference to 3D numerical simulations performed with FLAC3D code for different relative stiffness of the support, assuming plastic incompressibility for the ground ($\psi = 0$)

| $\frac{E}{E_s}$ | $\frac{K_s}{2G}$ | Numerical simulation with FLAC3D | | Classical CV-CF method | | Method of Bernaud-Rousset | | Method of Nguyen-Minh and Guo | | Empirical expression of De la Fuente et al. | |
|---|---|---|---|---|---|---|---|---|---|---|---|
| | | $\frac{2G}{\sigma_0}\frac{u_\infty}{R}$ | $\frac{p_{s\infty}}{\sigma_0}$ | $\frac{2G}{\sigma_0}\frac{u_\infty}{R}$ | $\frac{p_{s\infty}}{\sigma_0}$ | $\frac{2G}{\sigma_0}\frac{u_\infty}{R}$ | $\frac{p_{s\infty}}{\sigma_0}$ | $\frac{2G}{\sigma_0}\frac{u_\infty}{R}$ | $\frac{p_{s\infty}}{\sigma_0}$ | $\frac{2G}{\sigma_0}\frac{u_\infty}{R}$ | $\frac{p_{s\infty}}{\sigma_0}$ |
| 0.05 | 2.81 | 1.05 | 0.710 | 1.11 | 0.291 | 1.04 | 0.630 | 1.07 | 0.480 | 1.12 | 0.702 |
| 0.25 | 0.56 | 1.11 | 0.316 | 1.14 | 0.196 | 1.14 | 0.199 | 1.12 | 0.265 | 1.12 | 0.316 |
| 0.50 | 0.28 | 1.13 | 0.201 | 1.15 | 0.140 | 1.16 | 0.107 | 1.14 | 0.172 | 1.13 | 0.192 |
| 0.75 | 0.19 | 1.14 | 0.153 | 1.16 | 0.109 | 1.17 | 0.073 | 1.15 | 0.128 | 1.14 | 0.138 |
| 1 | 0.14 | 1.15 | 0.123 | 1.16 | 0.090 | 1.17 | 0.056 | 1.16 | 0.102 | 1.15 | 0.106 |

**Table 5.4** Comparison of the various implicit and empirical approaches with reference to 3D numerical simulations performed with FLAC3D code for different relative stiffness of the support assuming associate plasticity for the ground ($\psi = \phi$)

| $\frac{E}{E_s}$ | $\frac{K_s}{2G}$ | Numerical simulation with FLAC3D | | Classical CV-CF method | | Method of Bernaud-Rousset | | Method of Nguyen-Minh and Guo | | Empirical expression of De la Fuente et al. | |
|---|---|---|---|---|---|---|---|---|---|---|---|
| | | $\frac{2G}{\sigma_0}\frac{u_\infty}{R}$ | $\frac{p_\infty}{\sigma_0}$ | $\frac{2G}{\sigma_0}\frac{u_\infty}{R}$ | $\frac{p_\infty}{\sigma_0}$ | $\frac{2G}{\sigma_0}\frac{u_\infty}{R}$ | $\frac{p_\infty}{\sigma_0}$ | $\frac{2G}{\sigma_0}\frac{u_\infty}{R}$ | $\frac{p_\infty}{\sigma_0}$ | $\frac{2G}{\sigma_0}\frac{u_\infty}{R}$ | $\frac{p_\infty}{\sigma_0}$ |
| 0.05 | 2.81 | 1.30 | 0.710 | 1.31 | 0.246 | 1.15 | 0.658 | 1.24 | 0.424 | 1.327 | 0.699 |
| 0.25 | 0.56 | 1.35 | 0.318 | 1.34 | 0.194 | 1.32 | 0.227 | 1.29 | 0.286 | 1.334 | 0.337 |
| 0.50 | 0.28 | 1.37 | 0.218 | 1.36 | 0.154 | 1.37 | 0.126 | 1.33 | 0.207 | 1.343 | 0.224 |
| 0.75 | 0.19 | 1.36 | 0.176 | 1.37 | 0.128 | 1.39 | 0.087 | 1.35 | 0.110 | 1.352 | 0.180 |
| 1 | 0.14 | 1.41 | 0.148 | 1.38 | 0.110 | 1.40 | 0.066 | 1.37 | 0.134 | 1.361 | 0.158 |

For the cases considered in this example, we observe that the different approaches give close values for the displacement at equilibrium. The classical convergence-confinement method significantly underestimates the support pressure at equilibrium. A better estimate is obtained by the method of Nguyen-Minh and Guo. The empirical expression of De La Fuente et al. (2019) permits to obtain the support pressure with a good precision. In practice, when the design aims to limit the loads acting on the support and to ensure its stability, it is necessary to resort to methods allowing a sufficiently reliable estimate of the support pressure at equilibrium. The empirical expression of De La Fuente et al. (2019) is therefore very useful.

## 5.3 Taking into Account of a Pre-support or a Confining Pressure on the Face

Where underground structures are excavated at shallow depths in an environment where deformations need to be controlled or in poor soils, it is common to use pre-support techniques (pre-vaults, umbrella arches, jet grouting columns, face bolting) ahead of the advancing face.

### 5.3.1 Taking into Account of a Pre-support

The extension of the convergence-confinement method and in particular the evaluation of the deconfinement ratio and the displacement at the face taking into account the installation of a pre-support has been studied by Guilloux et al. (1996), from axisymmetric numerical simulations carried out with the finite element software CESAR-LCPC. A negative installation distance $d$ is considered for a pre-support like a pre-vault. The displacement at the face $u_{0,d}$ is interpolated between two limit values, the displacement at the face for an unsupported tunnel $u_{0,+\infty}$, and the displacement at the face in the case where the pre-support advances to an infinite distance ahead of the face $u_{0,-\infty}$. From a parametric study assuming that the ground remains elastic, these authors propose the following empirical formulas:

$$u_{0,-\infty} = u_0 \left( 0.74 - \frac{1}{(k_{sn} + 3)^{0.7}} \right) \frac{1}{k_{sn} + 1} \text{ avec } k_{sn} = \frac{K_s}{2G}$$

$$u_{0,d} = \frac{1}{\pi} \arctan\left( 2.8 \frac{d}{R} - 0.3 \right) \left( u_0 - u_{0,-\infty} \right) + \frac{1}{2} \left( u_0 + u_{0,-\infty} \right) \tag{5.22}$$

For a pre-support such as umbrella arch or jet-grouting columns, the authors recommend taking for $u_{0,d}$, the value $u_{0,-\infty}$.

Once the convergence at the face has been assessed, the convergence curve of the ground behind the face can be determined from the methods presented in Sects. 5.1 or 5.2.

### 5.3.2 Evaluation of the Deconfinement Ratio for TBM Excavation with a Pressurized Shield

For tunnels excavated with a pressurised shield, a confining pressure is applied in front of the TBM at the face and at the periphery of the skirt. The deconfinement ratio at the face is commonly estimated using the relation:

$$\lambda^* = \lambda_0 \left( 1 - \frac{P_{conf}}{\sigma_0} \right) \tag{5.23}$$

where $P_{conf}$ is the confining pressure acting of the face and $\lambda_0$ is the deconfinement rate at the face for unconfined face.

In practice, the use of 3D numerical simulations is essential for complex situations and the numerical tools available to the engineer today make these computations quite accessible.

## References

Bernaud D, Rousset G (1992) La "nouvelle méthode implicite" pour l'étude du dimensionnement des tunnels. Rev Fr Géotech 60:5–26

Bernaud D, Rousset G (1996) The 'new implicit method' for tunnel analysis. Int J Numer Anal Meth Geomech 20:673–690

Cantieni L, Anagnostou G (2009) The effect of the stress path on squeezing behavior in tunneling. Rock Mech Rock Eng 42:289–318

Corbetta F (1990) Nouvelles méthodes d'étude des tunnels profonds. Calculs nalytiques et numériques. Thèse de doctorat de l'Ecole Nationale Supérieure des Mines de Paris

Corbetta F, Bernaud D, Nguyen-Minh D (1991) Contribution à la méthode convergence-confinement par le principe de la similitude. Rev Fr Géotech 54:5–11

De La Fuente M (2018). Tunneling under squeezing conditions: effect of the excavation method. Thèse de doctorat de l'Université Paris-Est

De La Fuente M, Taherzadeh R, Sulem J, Nguyen X-S, Subrin D (2019) Applicability of the convergence-confinement method to full-face excavation of circular tunnels with stiff support system. Rock Mech Rock Eng 52(7):2361–2376

Guilloux A, Bretelle S, Bienvenue F (1996) Prise en compte des pré-soutènements dans le dimensionnement des tunnels. Rev Fr Géotech 76:3–16

Nguyen-Minh D, Guo C (1996) Recent progress in convergence-confinement method. In: Barla G (ed) Proceedings of Eurock'96, Prediction and performance in rock mechanics and rock engineering, Turin, Balkema, pp 855–860

Vlachopoulos N, Diederichs MS (2009) Improved longitudinal displacement profiles for convergence confinement analysis of deep tunnels. Rock Mech Rock Eng 42(2):131–146

# Chapter 6
# The Convergence-Confinement Method and the Time-Dependent Behavior of the Rock Mass

It has been shown in the previous chapters how the convergence of the ground and the support pressure increase with the distance from the tunnel face. But, for many structures, this growth cannot be explained only by the effect of the advance of the tunnel face, and it is necessary to take into account the time-dependent behavior of the ground and the phenomena of flow and consolidation in the case of tunnels excavated below the water table.

For example, during a long stop of the face advance, it has been observed that the convergence continues to grow. In addition, monitoring of ancient deep tunnels has shown that support pressures build up on linings which were installed far behind the face, long after the excavation period. These support pressures can only be explained by a slow creep/relaxation process in the rock mass which gradually induces the loading of the lining. A pronounced time-dependent behavior is particularly important in so-called 'squeezing' grounds characterized by high deformability and poor mechanical properties.

The prediction of long-term stresses in the supports thus requires the knowledge and the modeling of the time-dependent behavior of the rock.

## 6.1 Analysis of Convergence Measurements

Convergence measurements are the simplest and most direct way to understand the deformations of the rock mass induced by the excavation of an underground structure. The analysis of the convergence curves allows to differentiate between the effect of the advancing tunnel face and the effect of the time-dependent behavior of the rock mass by taking into account both the distance to the face $x$ and the time $t$ which has elapsed since the passage of the face to the considered section. This is usually done by fitting a convergence law in the form of a mathematical function. For this purpose, the following function has been proposed (Sulem 1983, Sulem et al. 1987a, b):

© The Author(s), under exclusive license to Springer Nature Switzerland AG 2022    123
M. Panet and J. Sulem, *Convergence-Confinement Method for Tunnel Design*,
Springer Tracts in Civil Engineering, https://doi.org/10.1007/978-3-030-93193-3_6

$$C(x, t) = C_{\infty x}\left[1 - \left(\frac{X}{x + X}\right)^2\right]\left\{1 + m\left[1 - \left(\frac{T}{t + T}\right)^n\right]\right\} \quad (6.1)$$

This convergence law depends on five parameters: $C_{\infty x}$ is the 'instantaneous' convergence corresponding to the convergence far from the distance of influence of the face in the absence of time-dependent effect, $X$ is a parameter related to the distance of influence of the face, $T$ is a characteristic time of the time-dependent response of the rock-support system, $m$ is a parameter giving the ratio between the instantaneous convergence and the ultimate total convergence (in the long-term, far from the distance of influence of the face) $C_\infty = C_{\infty x}(1 + m)$, and $n$ is an exponent, often taken equal to 0.3.

In this analysis, it must be taken into account that the monitoring targets are installed at a certain distance $x_0$ from the face and at time $t_0$, and that consequently the recorded convergence measurements correspond to:

$$\Delta C(x, t) = C(x, t) - C(x_0, t_0) \quad (6.2)$$

The method for fitting the parameters of the convergence law first uses the data corresponding to a stop of the advance of the face, allowing evaluating the parameters describing the time-dependent effects. For relatively homogeneous sections of a tunnel, we observe that the values of the parameters $X$, $T$ and $m$ can be fixed and that the fitting can then be done only on the sole parameter $C_{\infty x}$.

Table 6.1 presents some typical values of the parameters of the convergence law evaluated on different case studies.

The calibration of the convergence law permits an evaluation by inverse analysis of the mechanical parameters of the rock mass. In particular, it makes it possible to distinguish between short-term and long-term parameters. For example, the parameter $C_{\infty x}$ that represents the convergence of the instantaneously excavated tunnel can be used to evaluate the elastoplastic parameters such as the shear modulus and the cohesion of the rock mass. This method permits to overcome the limitations of extrapolating parameters measured in the laboratory to field conditions.

**Table 6.1** Parameters of the convergence law evaluated on few case studies

| Tunnel | Type of rock mass | X/D | m | T (days) |
|---|---|---|---|---|
| Fréjus | Calcschists | 0.9 | 4.5 | 0.5–5 |
| Las Planas | Marls | 0.45 | 3 | 2.3 |
| Chamoise | Jurassic marls | 0.54 | 4.3 | 2.5 |
| Monaco | Cenomanian marls | 0.45 | 1.6 | 24.5 |
| Underground Research Laboratory of Meuse/Haute-Marne | Argilite | 0.9 | 5.7 | 6 |
| St Martin-la-Porte access gallery | Carboniferous schists | 1.25 | 9–18 | 15–90 |

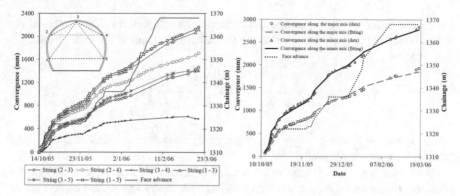

**Fig. 6.1** Example of Saint-Martin-la-Porte access gallery. Chainage PM 1311. **a** Measured convergence along five strings; **b** Convergence curves along the principal axes of deformation and comparison between measured and fitted values ($T = 70.5$ days; $m = 9.5$; $X = 13.4$ m; major axis: $C_{\infty x} = 0.50$ m; minor axis: $C_{\infty x} = 0.75$ m), (after Tran-Manh et al. 2015)

In squeezing grounds, strong anisotropic deformation is commonly observed around the tunnel, leading to an ovalization of the section. Convergence measurements in different directions can be used to study the transition from a quasi-circular to an elliptical section (Vu et al. 2013; Guayacán-Carrillo et al. 2016a, b). The orientation of the ellipse, which generally stabilizes after the first convergence measurements, gives the directions of principal deformation. The convergence law can be used to calculate the convergences along the major axis (direction of the smallest convergence) and along the minor axis (direction of the strongest convergence) (Fig. 6.1).

## 6.2 Excavation of Tunnel with a Circular Section in a Viscoelastic Rock Mass

Let's consider a viscoelastic rock mass whose behavior is described by a Kelvin-Voigt model, (Fig. 6.2). It is assumed that the volume behavior of the rock mass is purely

**Fig. 6.2** Kelvin-Voigt viscoelastic model

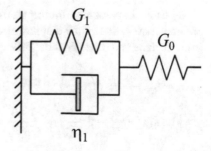

elastic and that only the deviatoric deformations are viscoelastic. The stress–strain relationship for the deviatoric part of the stress tensor ($s$) and of the strain tensor ($e$) is written as:

$$\left(\frac{1}{T_0} + \frac{d}{dt}\right)s_{ij} = 2G_0\left(\frac{1}{T_1} + \frac{d}{dt}\right)e_{ij} \tag{6.3}$$

where $T_0 = \frac{\eta_1}{G_0 + G_1}$ and $T_1 = \frac{\eta_1}{G_1}$ are respectively the relaxation time and the creep time of the viscoelastic model.

For a tunnel with a circular cross-section of radius $R$ excavated in a homogenous, elastic, isotropic rock mass under an isotropic initial stress state $\sigma_0$, the radial displacement $u_R$ at the wall is given by (see Eq. 3.11):

$$\frac{u_R(x)}{R} = \lambda(x)\frac{\sigma_0}{2G} \tag{6.4}$$

where $x$ is the distance of the considered section from the face, $\lambda(x)$ is the corresponding deconfinement ratio, and $G$ the elastic shear modulus of the medium.

This solution can be extended to the viscoelastic case by using the relation (6.3):

$$2G_0\left(\frac{1}{T_1} + \frac{d}{dt}\right)\left(\frac{u_R}{R}\right) = \left(\frac{1}{T_0} + \frac{d}{dt}\right)(\lambda\sigma_0) \tag{6.5}$$

A closed-form solution can be obtained by assuming a deconfinement ratio that can be described by an exponential function:

$$\lambda(x) = \lambda_0 + (1 - \lambda_0)(1 - \exp(-x/X)) \tag{6.6}$$

where $X$ is a parameter related to the distance of influence of the advancing face.

If we assume that the front advances at a constant rate $V_a$ ($x = V_a t$), the evolution in time of the deconfinement ratio is written as:

$$\lambda(x) = \lambda_0 + (1 - \lambda_0)(1 - \exp(-t/T_a)), \ \ T_a = \frac{X}{V_a} \tag{6.7}$$

By using Laplace-Carson transform, one can solve the differential Eq. (6.5) to obtain the evolution of the displacement at the tunnel wall as a function of distance from the face and time (Panet 1979):

$$u_R = u_\infty - A\exp\left(-\frac{x}{X}\right) - B\exp\left(-\frac{t}{T}\right)$$

$$\text{with}: u_\infty = \frac{\sigma_0 R}{2G_\infty}; \ \ \frac{1}{G_\infty} = \frac{1}{G_0}\frac{T_1}{T_0} = \frac{1}{G_0} + \frac{1}{G_1}$$

and : $A = \dfrac{\sigma_0 R}{2G_0}(1 - \lambda_0)\dfrac{T_1/T_a - T_1/T_0}{T_1/T_a - 1}$; $B = \dfrac{\sigma_0 R}{2G_0}(T_1/T_0 - 1)\left(\dfrac{T/T_a - \lambda_0}{T/T_a - 1}\right)$

For $x = 0$ and $t = 0$, $u_R = u_\infty - A - B = \lambda_0 \dfrac{\sigma_0 R}{2G_0}$ (6.8)

In the above solution, it is assumed that the time-dependent deformations that occur before the passage of the face can be neglected.

We can define a convergence curve of the unsupported ground in the short term corresponding to an elastic response and given by the equation:

$$\sigma_R + 2G_0 \frac{u_R}{R} - \sigma_0 = 0 \qquad (6.9)$$

and a convergence curve of the unsupported ground in the long term

$$\sigma_R + 2G_\infty \frac{u_R}{R} - \sigma_0 = 0 \qquad (6.10)$$

The convergence confinement method can be applied to obtain the short term support pressure $p_{s0}$, i.e. immediately after construction by the intersection of the short term convergence curve and the support confinement curve (SCC):

$$p_{s0} = \frac{K_{sn}}{K_{sn} + 2G_0}(1 - \lambda_d)\sigma_0 \qquad (6.11)$$

and the long term support pressure $p_{s\infty}$, by the intersection of the long term convergence curve, and the support confinement curve:

$$p_{s\infty} = \frac{K_{sn}}{K_{sn} + 2G_\infty}\left(1 - \frac{G_\infty}{G_0}\lambda_d\right)\sigma_0 \qquad (6.12)$$

where $\lambda_d$ is the deconfinement ratio at the time of the support installation (Fig. 6.3).

It is important to note that in the above expressions, the excavation of the ground is assumed to be sufficiently rapid so that the radial displacement at the support installation distance $u_d$ can be assumed to be purely elastic, $u_d = \lambda_d \frac{\sigma_0}{2G_0} R$. Moreover, the creep of the support is not considered here.

## 6.3 Excavation of Tunnel with a Circular Section in a Visco-Elastic–plastic Rock Mass

In order to account for the plastic behavior of the rock mass, the Kelvin-Voigt viscoelastic model can be extended by assembling in series a Kelvin viscoelastic

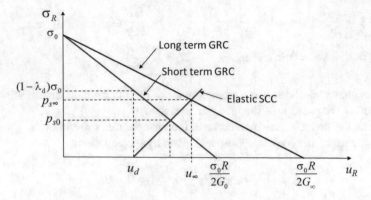

**Fig. 6.3** Application of the convergence-confinement method in a viscoelastic medium

element, a Maxwell viscoelastic element and a Mohr–Coulomb plastic element (Fig. 6.4). This model is known as the 'CVISC' model. The volumetric deformations are assumed to be elasto-plastic and the deviatoric deformations, elasto-visco-plastic:

$$\varepsilon_{ij} = e_{ij} + \varepsilon_v \delta_{ij}/3, \quad \sigma_{ij} = s_{ij} + \sigma_{kk}\delta_{ij}/3$$
$$e_{ij} = e_{ij}^K + e_{ij}^M + e_{ij}^p$$

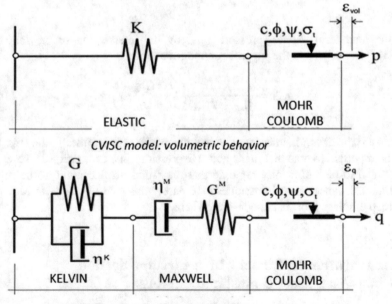

**Fig. 6.4** CVISC visco-elasto-plastic model

$$\text{Kelvin element: } s_{ij} = 2G^K e_{ij}^K + 2\eta^K \dot{e}_{ij}^K$$

$$\text{Maxwell element: } \dot{e}_{ij}^M = \frac{\dot{s}_{ij}}{2G^M} + \frac{s_{ij}}{2\eta^M}$$

$$\text{Plastic flow rule : } \dot{\varepsilon}_{ij}^p = \dot{\psi} \frac{\partial Q}{\partial \sigma_{ij}} \tag{6.13}$$

For the plastic element, we consider a Mohr–Coulomb yield criterion $F$ and plastic potential $G$ that can be expressed in terms of principal stresses $(\sigma_1, \sigma_3)$ as:

$$\text{Yield criterion: } F = \sigma_1 - K_p \sigma_3 - 2C\sqrt{K_p}, \quad K_p = \frac{1 + \sin\phi}{1 - \sin\phi}$$

$$\text{Plastic potential: } Q = \sigma_1 - K_\psi \sigma_3, \quad K_\psi = \frac{1 + \sin\psi}{1 - \sin\psi} \tag{6.14}$$

where $\phi$ is the internal friction angle of the material, $C$ is the cohesion and $\psi$ is the dilatancy angle.

Tran-Manh et al. (2015) have proposed an analytical solution for the stress and displacement fields around a tunnel with circular cross section in the form:

*Stress field*

In the visco - elasto - plastic zone: $R \leq r \leq R_p$

$$\sigma_r = \frac{2C\sqrt{K_p}}{K_p - 1}\left[\left(\frac{r}{R}\right)^{K_p - 1} - 1\right] + (1 - \lambda)\sigma_0\left(\frac{r}{R}\right)^{K_p - 1}$$

$$\sigma_\theta = \frac{2C\sqrt{K_p}}{K_p - 1}\left[K_p\left(\frac{r}{R}\right)^{K_p - 1} - 1\right] + K_p(1 - \lambda)\sigma_0\left(\frac{r}{R}\right)^{K_p - 1}$$

$$\text{Plastic radius: } \frac{R_p}{R} = \left(\frac{2}{K_p + 1}\frac{(K_p - 1)\sigma_0 + 2C\sqrt{K_p}}{(1 - \lambda)(K_p - 1)\sigma_0 + 2c\sqrt{K_p}}\right)^{\frac{1}{K_p - 1}}$$

In the visco - elastic zone: $r \geq R_p$

$$\sigma_r = \left(1 - \lambda_e\left(\frac{R_p}{r}\right)^2\right)\sigma_0$$

$$\sigma_\theta = \left(1 + \lambda_e\left(\frac{R_p}{r}\right)^2\right)\sigma_0$$

$$\lambda_e = \frac{1}{K_p + 1}\left(K_p - 1 + \frac{2C\sqrt{K_p}}{\sigma_0}\right) \tag{6.15}$$

*Displacement field*

In the visco-elasto-plastic zone: $R \leq r \leq R_p$

$$\frac{\Delta u}{r} = \frac{D}{K_p + K_\psi}\left(\frac{r}{R}\right)^{K_p-1} + H\left(\frac{R_p}{r}\right)^{K_\psi+1}$$

$$D = \frac{1}{2}(1 - K_\psi)\left[(1 - K_p)(1 - \lambda)\sigma_0 - 2C\sqrt{K_p}\right]$$

$$\left(\frac{t}{2\eta^M} + \frac{1}{2G^M} + \frac{1}{2G^K}\left[1 - \exp\left(-\frac{t}{T^K}\right)\right]\right)$$

$$H = \left(\lambda_e\sigma_0 - \frac{(1 - K_\psi)\left[(1 - K_p)(1 - \lambda)\sigma_0 - 2C\sqrt{K_p}\right]}{2(K_p + K_\psi)}\left(\frac{R_p}{R}\right)^{K_p-1}\right)$$

$$\times \left(\frac{t}{2\eta^M} + \frac{1}{2G^M} + \frac{1}{2G^K}\left[1 - \exp\left(-\frac{t}{T^K}\right)\right]\right) \tag{6.16}$$

Note that for this constitutive model, the stress field and the radius of the plastic zone are the same as for an elastoplastic model without viscosity. Once the tunnel is fully excavated ($\lambda = 1$), the plastic radius does not evolve in time.

## 6.4   Viscoplastic Models

*Bingham model*

Although viscoelastic models are widely used in practice, the time-dependent behavior of some rocks such as salt rocks or clayey rocks is better represented by a viscoplastic model. The simplest viscoplastic model is the Bingham model which associates in series an elastic spring (Hooke's model) which represents the instantaneous elastic deformations and sliding frictional element in parallel with a viscous element which represents the time-dependent viscoplastic deformations (Fig. 6.5).

**Fig. 6.5** Bingham viscoplastic model

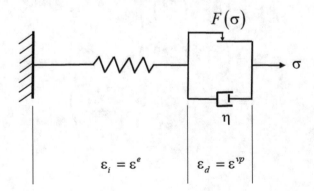

The strain rate $\dot{\varepsilon}$ is the sum of the elastic strain rate $\dot{\varepsilon}^e$ and of the viscoplastic strain rate $\dot{\varepsilon}^{vp}$:

$$\dot{\varepsilon} = \dot{\varepsilon}^e + \dot{\varepsilon}^{vp} \tag{6.17}$$

The viscoplastic strain rate is expressed in the general form (Perzyna, 1966):

$$\dot{\varepsilon}_{ij}^{vp} = \frac{1}{\eta} \langle F(\sigma) \rangle^n \frac{\partial Q(\sigma)}{\partial \sigma_{ij}} \tag{6.18}$$

where:

> $\eta$ is the viscosity coefficient
>
> $n$ is a constant
>
> $F(\sigma)$ is the yield function of the viscoplastic model
>
> $Q(\sigma)$ is the plastic potential
>
> $\langle F(\sigma) \rangle$ is the positive part of $F(\sigma)$.

Plane strain stress–strain relationships are written as:

$$\begin{aligned} \dot{\varepsilon}_r &= \frac{\partial \dot{u}}{\partial r} = \frac{1-\nu}{2G}\dot{\sigma}_r - \frac{\nu}{2G}\dot{\sigma}_\theta + 2\dot{\varepsilon}_r^{vp} \\ \dot{\varepsilon}_\theta &= \frac{\dot{u}}{r} = \frac{-\nu}{2G}\dot{\sigma}_r + \frac{1-\nu}{2G}\dot{\sigma}_\theta - 2\dot{\varepsilon}_\theta^{vp} \end{aligned} \tag{6.19}$$

By taking into account the equilibrium equation

$$\frac{\partial \dot{\sigma}_r}{\partial r} = \frac{\dot{\sigma}_\theta - \dot{\sigma}_r}{r} \tag{6.20}$$

and the dilatancy flow rule

$$\dot{\varepsilon}_r^{vp} + K_\psi \dot{\varepsilon}_\theta^{vp} = 0 \tag{6.21}$$

the following differential equation is obtained:

$$\frac{\partial \dot{u}}{\partial r} + K_\psi \frac{\dot{u}}{r} = \frac{1}{2G}\left( (1-2\nu)(K_\psi + 1)\dot{\sigma}_r + \left(K_\psi - (K_\psi + 1)\nu\right)r\frac{\partial \dot{\sigma}_r}{\partial r} \right) \tag{6.22}$$

Bérest and Nguyen-Minh (1983) have studied the effect of excavation and support installation for a viscoplastic rock mass under the assumption of plastic incompressibility ($K_\psi = 1$), and with a linear Tresca yield criterion ($n = 1$).

Boundary conditions express the radial stress conditions imposed at infinity and at the tunnel wall represented by the deconfinement ratio $\lambda$ as well as the continuity of the displacements and the equilibrium of the stress vectors at the boundary between the elastic zone and the viscoplastic zone. The extent of the viscoplastic zone is defined by the viscoplastic radius $R_{vp}$ which evolves with time.

Assuming plastic incompressibility ($K_\psi = 1$), Eq. (6.22) becomes:

$$\frac{\partial \dot{u}}{\partial r} + \frac{\dot{u}}{r} = \frac{(1 - 2v)}{2G}\left(2\dot{\sigma}_r + r\frac{\partial \dot{\sigma}_r}{\partial r}\right) \tag{6.23}$$

Boundary conditions are written as:

$$\begin{aligned} r = R &: \ \dot{\sigma}_r = \dot{p}_i \\ r = R_{vp} &: \ \dot{u} = \lambda_e \frac{\sigma_0}{2G}\dot{R}_{vp} \end{aligned} \tag{6.24}$$

We obtain:

$$2G\frac{\dot{u}}{r} = (1 - 2v)\dot{\sigma}_r + 2(1 - v)\lambda_e\frac{\dot{R}_{vp}^2}{r^2} \tag{6.25}$$

where $\lambda_e$ is the deconfinement ratio at the onset of plasticity and is defined by: $F((1 - \lambda_e)\sigma_0, (1 + \lambda_e)\sigma_0) = 0$.

For $r = R$, Eq. (6.25) becomes:

$$2G\frac{\dot{u}_R}{R} = (1 - 2v)\dot{\sigma}_R + 2(1 - v)\lambda_e\frac{\dot{R}_{vp}^2}{R^2} \tag{6.26}$$

If an elastic lining of normal stiffness $K_{sn}$ is installed, the support pressure evolves in time and its derivative with respect to time is written as:

$$\dot{p}_s = K_{sn}\dot{u}_R \tag{6.27}$$

Using equation, we obtain:

$$\left(2G - (1 - 2v)K_{sn}\right)\frac{\dot{u}_R}{R} = 2(1 - v)\lambda_e\frac{\dot{R}_{vp}^2}{R^2} \tag{6.28}$$

As $\dot{u}_R$ is positive, $\frac{d}{dt}R_{vp}^2$ has the sign of $(2G - (1 - 2v)K_{sn})$.

If $K_{sn} < G/(1 - 2v)$, the support is said to be soft and the viscoplastic radius $R_{vp}$ continuously increases, whereas in the opposite case, the support is said to be stiff and $R_{vp}$ decreases when the support is installed, which means that a zone inside the rock mass undergoes an elastic unloading. Let $R_{vp}^{\max}$ denote the maximum value reached by the plastic radius during the excavation Three zones can be distinguished in the rock mass:

- an elastic zone:for $r > R_{vp}$, $\varepsilon_{vp} = 0$ and $\dot{\varepsilon}_{vp} = 0$
- an elastic zone with residual plastic deformations: for $R_{vp} < r \le R_{vp}^{\max}$, $\varepsilon_{vp} \ne 0$ and $\dot{\varepsilon}_{vp} = 0$
- a viscoplastic zone: for $R \le r \le R_{vp}$, $\varepsilon_{vp} \ne 0$ and $\dot{\varepsilon}_{vp} \ne 0$.

*SHELVIP model*

The viscoplastic model *SHELVIP* (Stress Hardening ELastic VIscous Plastic model) was proposed by Debernardi and Barla, (2009) for excavations in severe squeezing conditions. It couples a viscoplastic Bingham element, and a plastic element:

$$\dot{\varepsilon} = \dot{\varepsilon}^e + \dot{\varepsilon}^p + \dot{\varepsilon}^{vp} \tag{6.29}$$

These authors have proposed analytical expressions to describe the creep and relaxation tests. The implementation in computational software has shown that this type of model allows to reproduce well the strong convergences observed in tunnels excavated in squeezing ground (Barla et al. 2010).

## 6.5 Taking into Account the Hydraulic Regime

The excavation of a tunnel located under the water table, in a saturated rock mass, modifies the hydraulic conditions around the tunnel and thus the pore pressure field. Taking into account the sequence of tunnel excavation and lining installation in the evolution of the pressure field generally requires the use of numerical models. However, under certain assumptions, closed-form solutions which are useful for preliminary design can be obtained. Bobet (2010) has proposed an analytical solution for the stress and pore pressure fields for a circular tunnel excavated in an isotropic, homogeneous and saturated medium with an elasto-plastic behavior. These solutions allow to establish the short and long term ground characteristics curves for different permeability conditions of the lining.

### 6.5.1 Ground Reaction Curve for a Poroelastic Medium

The reader can find a presentation of the fundamental concepts of poromechanics in the book by Coussy (2004).

For a poroelastic medium, the constitutive relations are written as a function of the Biot effective stresses:

$$\sigma_r' = \sigma_r - bp_f = (\lambda + 2G)\frac{\partial u}{\partial r} + \lambda\frac{u}{r}$$

$$\sigma_\theta' = \sigma_\theta - bp_f = \lambda\frac{\partial u}{\partial r} + (\lambda + 2G)\frac{u}{r} \tag{6.30}$$

where $p_f$ is the pore pressure and $b$ is the Biot coefficient.

In axi-symmetric conditions, the equilibrium equation is written as:

$$\frac{\partial \dot{\sigma}_r}{\partial r} = \frac{\dot{\sigma}_\theta - \dot{\sigma}_r}{r} \tag{6.31}$$

The balance equation for the fluid mass combined with Darcy's law gives the following equation for the evolution of the fluid mass per unit volume of the porous medium $m_f$:

$$\frac{\partial m_f}{\partial t} = \rho_f \frac{k}{\eta} \nabla^2 p_f \tag{6.32}$$

where $k$ is the intrinsic permeability of the medium, $\rho_f$ is the density of the fluid and $\eta$ is its kinematic viscosity.

The boundary conditions express that far from the tunnel wall, the stresses and the pore pressure remain unchanged and that at the tunnel wall, the radial stress is imposed with an additional hydraulic boundary condition corresponding either to an imposed pore pressure (drained case) or to zero fluid flux (undrained case):

$$\text{for } r \to \infty, \begin{cases} \sigma_r = \sigma_0 \\ p_f = p_0 \end{cases}$$

$$\text{for } r = R, \begin{cases} \sigma_r = \sigma_i \\ p_f = p_i \text{ (drained case) or } \frac{\partial p_f}{\partial r} = 0 \text{ (undrained case)} \end{cases} \tag{6.33}$$

In the general case, the diffusion Eq. (6.32) must be solved numerically to obtain the transient solution. Two extreme cases are of particular interest: the short term response corresponding to a period of time close enough to the excavation so that on can consider that the overpressures have not yet dissipated, and the long term response for which the steady state is reached ($\partial/\partial t = 0$).

For the steady state, Eq. (6.32) becomes:

$$\nabla^2 p_f = 0 \tag{6.34}$$

Assuming an undrained boundary condition at the tunnel wall, corresponding to an impervious lining or a ground of very low permeability, the solution of Eq. (6.34) corresponds to a pore pressure which remains unchanged:

$$p_f = p_0 \tag{6.35}$$

We deduce the stress and displacement fields in the undrained regime:

$$\sigma_r = \sigma_0 - (\sigma_0 - \sigma_i)\left(\frac{R}{r}\right)^2$$

$$\sigma_\theta = \sigma_0 + (\sigma_0 - \sigma_i)\left(\frac{R}{r}\right)^2$$

$$2G\frac{\Delta u}{r} = (\sigma_0 - \sigma_i)\left(\frac{R}{r}\right)^2$$

$$\text{at the wall, } r = R, \quad \frac{\Delta u_R}{R} = \frac{1}{2G}(\sigma_0 - \sigma_i)$$

$$\sigma_\theta - \sigma_r = 2(\sigma_0 - \sigma_i) \tag{6.36}$$

It should be noted that in the undrained condition, the total stress and displacement fields are identical to the case of a tunnel excavated in a dry elastic medium.

For a drained condition at the wall, it is assumed that beyond a sufficiently large distance from the wall $R_d$, the fluid pressure remains unchanged. The solution of Eq. (6.34) corresponds to a logarithmic evolution of the pore pressure in the vicinity of the excavation:

$$p_f = p_i + \frac{p_0 - p_i}{\ln \frac{R_d}{R}} \ln \frac{r}{R} \tag{6.37}$$

We deduce the stress and displacement fields in the drained regime:

$$\sigma_r = \sigma_0 - (\sigma_0 - \sigma_i)\left(\frac{R}{r}\right)^2 + \frac{1-2v}{2(1-v)}b(p_0 - p_i)\left[\left(\frac{R}{r}\right)^2 + \frac{\ln \frac{r}{R_d}}{\ln \frac{R_d}{R}}\right]$$

$$\sigma_\theta = \sigma_0 + (\sigma_0 - \sigma_i)\left(\frac{R}{r}\right)^2 - \frac{1-2v}{2(1-v)}b(p_0 - p_i)\left[\left(\frac{R}{r}\right)^2 - \frac{\ln \frac{r}{R_d}}{\ln \frac{R_d}{R}}\right]$$

$$\sigma_x = \sigma_0 + \frac{1-2v}{1-v}b(p_0 - p_i)\frac{\ln \frac{r}{R_d}}{\ln \frac{R_d}{R}}$$

$$2G\frac{\Delta u}{r} = (\sigma_0 - \sigma_i)\left(\frac{R}{r}\right)^2 - \frac{1-2v}{2(1-v)}b(p_0 - p_i)\frac{1}{\ln \frac{R_d}{R}}\left(\left(\frac{R}{r}\right)^2 + \ln \frac{r}{R_d}\right)$$

$$\text{at the wall, } r = R, \; R_d \gg R, \quad \frac{\Delta u}{R} = \frac{1}{2G}(\sigma_0 - \sigma_1) + \frac{1}{2G}\frac{1-2v}{2(1-v)}b(p_0 - p_i)$$

$$\sigma_\theta - \sigma_r = 2(\sigma_0 - \sigma_1) - \frac{1-2v}{2(1-v)}b(p_0 - p_i) \tag{6.38}$$

It should be noted that in drained conditions, the total stress along the tunnel axis $\sigma_x$ is affected by the change in pore pressure, and the displacement at the wall is all the greater as the interstitial pressure drop at the wall is strong.

### 6.5.2 Application of the Convergence-Confinement Method for a Poroelastic Saturated Medium

Let's consider an elastic support of normal stiffness $K_{sn}$ installed at a distance $d$ from the face, corresponding to a radial displacement $u_d$:

Support confining curve (SCC) :

$$p_s = K_{sn}\left(\frac{u_R}{R} - \frac{u_d}{R}\right)$$

Ground reaction curve (GRC) :

Undrained (short term): $\sigma_i = \sigma_0 - 2G\frac{u_R}{R}$

Drained (long term): $\sigma_i = \sigma_0 - 2G\frac{u_R}{R} + \frac{1-2v}{2(1-v)}b(p_0 - p_i)$     (6.39)

If it is assumed that the ground is of very low permeability so that the excavation can be considered to be performed in undrained conditions, $u_d = \lambda_d \frac{\sigma_0}{2G}$.

We distinguish the case of an impervious lining for which the support pressure at equilibrium is given by:

$$\text{Impervious lining: } p_s = \frac{K_{sn}}{K_{sn} + 2G}\sigma_0(1 - \lambda_d)$$     (6.40)

and the case of a pervious lining ($p_i = 0$) for which the long term support pressure is:

$$\text{Pervious lining: } p_s = \frac{K_{sn}}{K_{sn} + 2G}\left(\sigma_0(1 - \lambda_d) + \frac{1-2v}{2(1-v)}bp_0\right)$$     (6.41)

### 6.5.3 Time to Establish the Steady-State Hydraulic Regime

In the previous sections, we have presented solutions in the steady-state hydraulic regime. To reach this regime, it is necessary to have an equipotential surface through which a permanent recharge occurs. If not, the flow of water tends asymptotically to zero and the ground progressively desaturates. The time for the establishment of this permanent regime can be estimated as (Rat 1973):

$$\text{Case of an infinite cyclinder: } t_f = 2\frac{S_s}{k_w}L_w^2$$

$$\text{Case of a sphere } t_f = \frac{4}{\pi}\frac{S_s}{k_w}L_w^2$$     (6.42)

where

$L_w$ is the distance between the axis of the tunnel and the equipotential surface ($L_w \gg R$)

$k_w = \dfrac{\rho_f g}{\eta} k$ is the coefficient of hydraulic diffusivity

$S_s = \rho_f \left( \dfrac{1}{K_d} + \dfrac{n}{K_f} \right)$ is the storage coefficient

$K_d$ is the drained bulk modulus of the solid skeleton

$K_f$ is the bulk modulus of the fluid

$n$ is the porosity of the medium (6.43)

The case of the infinite cylinder can be used if the tunnel lining can be considered as permeable with respect to the permeability of the ground, and the case of the sphere when the lining has a low permeability with respect to that of the ground.

The expression of the storage coefficient $S_s$ given above and evaluated from poroelasticity theory corresponds to the case of a captive groundwater table. For free groundwater tables, we define the specific storage coefficient ($S_s \times L$ where $L$ is the thickness of the aquifer) which is approximately equal to the porosity of the soil (gravel and sand).

In practice, this time $t_f$ should be compared with the time $t_a$ which characterizes the rate of excavation. This time can vary considerably depending on the ground encountered. It is a few hours in very permeable soils, a few days in the Cenomanian blue chalk of the Channel Tunnel, and several years in very low permeability soils such as clay. In many situations, $t_f \gg t_a$, which justifies considering that the excavation corresponds to undrained conditions.

By taking into account the transient effects, it is shown that for a very low permeability ground, transient pore pressures can be generated in the vicinity of the tunnel face, during the excavation. This phenomenon, which reflects the hydro-mechanical coupling, is known as the Mandel-Cryer effect (Mandel 1953; Cryer 1963; Detournay and Cheng 1993; Coussy 2004). As the pore pressure begins to dissipate at the wall, there is a redistribution of stresses within the rock mass leading to the generation of a local pore overpressure that may take some time to dissipate due to the low permeability of the ground.

### 6.5.4 Excavation in a Saturated Poroplastic Medium

Taking into account the hydromechanical coupling, in the case of an elastoplastic behavior of the ground, requires much more complex developments as done in the works of Giraud et al. (1993), Benamar (1996), Giraud et al. (2002). After the excavation considered as quasi instantaneous, the dilatant behavior of the ground in the plastic zone leads to a non-uniform pore pressure drop which results in a flow of the pore fluid. The evolving pore-pressure induces an evolution of the state of stress in the rock before reaching the long term conditions. This evolution depends on the behavior of the ground and the drainage conditions at the wall. In the case of a pervious lining, the delayed convergences remain weak for a Tresca or Mohr–Coulomb plasticity criterion. On the other hand, these delayed convergences can be important with an impervious lining in the case of a Mohr–Coulomb criterion with a risk of full collapse of the tunnel. Giraud et al. (1993) have shown that the evolution of the plastic zone differs in the two cases. For a pervious lining (drained conditions at the wall), the plastic radius decreases after a certain time and the response of the rock mass becomes elastic, which explains why the delayed convergence remains limited. On the other hand, for an impervious lining (undrained conditions at the wall), the evolution of the plastic zone is more complex and non-monotonous. These theoretical and numerical developments show the necessity of the monitoring and the collection of field data which must be carried out on the structures during long periods.

# References

Barla G, Bonini M, Debernardi D (2010) Time dependent deformations in squeezing tunnels. Int J Geoeng Case Hist 2(1):40–65

Benamar I (1996) Etude des effets différés dans les tunnels profonds. Thèse de doctorat de l' Ecole Nationale des Ponts et Chaussées

Bérest P, Nguyen-Minh D (1983) Modèle viscoplastique pour le comportement d'un tunnel revêtu. Rev Fr Géotech 23:19–25

Bernaud D (1991) Tunnels profonds dans les milieux viscoplastiques: Approches expérimentale et numérique. Thèse de doctorat de l'Ecole Nationale des Ponts et Chaussées

Bobet A (2010) Characteristic curves for deep circular tunnels in poroplastic rock. Rock Mech Rock Eng 43(2):185–200

Coussy O (2004) Poromechanics. Wiley

Cryer CW (1963) A comparison of the three-dimensional consolidation theories of Biot and Terzaghi. Quart J Mech Appl Math 16:401–412

De la Fuente M, Sulem J, Taherzadeh R, Subrin D (2020) Tunneling in squeezing ground: effect of the excavation method. Rock Mech Rock Eng 53(2):601–623

De La Fuente M (2018) Tunneling under squeezing conditions: effect of the excavation method. Thèse de doctorat de l'Université Paris-Est

Debernardi D, Barla G (2009) New viscoplastic model for design analysis of tunnels in squeezing conditions. Rock Mech Rock Eng 42(2):259–288

Detournay E, Cheng AH (1993) Fundamentals of poroelasticity. In: Fairhurst C (ed) Comprehensive rock engineering, Vol. 2, Analysis and design method. Pergamon Press, pp 113–171

Giraud A, Picard JM, Rousset G (1993) Time Dependent behavior of tunnels excavated in porous mass. Int J Rock Mech Min Sci Geomech Abstracts 30(7):1453–1459

Giraud A, Homand F, Labiouse V (2002) Explicit solutions for the instantaneous undrained contraction of hollow cylinders and spheres in porous elastoplastic medium. Int J Numer Anal Meth Geomech 26(3):231–258

Giraud A (1993) Couplages thermo-hydro-mécaniques dans les milieux poreux peu perméables: application aux argiles profondes. Thèse de doctorat de l'Ecole Nationale des Ponts et Chaussées

Guayacán-Carrillo L-M, Sulem J, Seyedi DM, Ghabezloo S, Noiret A, Armand G (2016) Convergence analysis of an unsupported micro-tunnel at the Meuse/Haute-Marne underground research laboratory. Geol Soc London Spec Publ 443(SP443):24

Guayacán-Carrillo L-M, Sulem J, Seyedi DM, Ghabezloo S, Noiret A, Armand G (2016) Analysis of long-term anisotropic convergence in drifts excavated in callovo-oxfordian claystone. Rock Mech Rock Eng 49(1):97–114

Guayacán-Carrillo L-M (2016) Analysis of long-term closure in drifts excavated in Callovo-Oxfordian claystone: roles of anisotropy and hydromechanical couplings. Thèse de doctorat de l'Université Paris-Est

Ladanyi B (1993) Time dependent response around tunnels. In: Fairhurst C (ed) Comprehensive rock engineering, Vol. 2, Analysis and design method. Pergamon Press, pp 77–112

Li X (1999) Stress and displacement fields around a deep circular tunnel with partial sealing. Comput Geotech 24(2):125–140

Mandel J (1953) Consolidation des sols (étude mathématique). Géotechnique 3:287–299

Panet M (1979) Les déformations différées dans les ouvrages souterrains. In: Procedings of 4th international congress on international society for rock mechanics, vol 3, Zurich, pp 291–303

Panet M (1993) Understanding deformations in tunnels. In: Comprehensive rock engineering, vol 1, 27. Pergamon Press, London, pp 663–690

Perzyna P (1966) Fundamental problems in viscoplasticity. Adv Appl Mech 9:243–377

Rat M (1973) Ecoulement et répartition des pressions interstitielles autour des tunnels. Bull Liaison Labo Ponts Et Chaussées 68:109–124

Rousset G (1990) Les sollicitations à long terme des revêtements des tunnels. Rev Fr Géotech 53:5–20

Rousset G (1988) Comportement mécanique des argiles profondes. Application au stockage de déchets radioactifs. Thèse de doctorat de l'Ecole Nationale des Ponts et Chaussées

Sakuraï S (1978) Approximate time-dependent analysis of tunnel support strcutures considering progress of tunnel face. Int J Num Anal Meth Geomech 2(2):159–175

Samarasekera L, Eisenstein Z (1992) Pore pressures aronnd tunnels in clay. Can Geotech J 29:819–831

Sulem J, Panet M, Guenot A (1987a) Closure analysis in deep tunnels. Int J Rock Mech Min Sci Geomech Abstr 24(3):145–154

Sulem J, Panet M, Guenot A (1987b) An analytical solution for time-dependent displacements in a circular tunnel. Int J Rock Mech Min Sci Geomech Abstr 24(3):155–164

Sulem J (1983) Comportement différé des galeries profondes. Thèse de doctorat de l'Ecole Nationale des Ponts et Chaussées

Tran-Manh H, Sulem J, Subrin D, Billaux D (2015) Anisotropic time-dependent modeling of tunnel excavation in squeezing ground. Rock Mech Rock Eng 48(6):2301–2317

Tran-Manh H (2014) Comportement des tunnels en terrain poussant. Thèse de doctorat de l'Université Paris-Est

Vu TM, Sulem J, Subrin D, Monin N, Lascols J (2013) Anisotropic closure in squeezing rocks: the example of Saint-Martin-la-Porte access gallery. Rock Mech Rock Eng 46(2):231–246

# Chapter 7
# Use of Numerical Models

## 7.1 Background

Analytical solutions are a valuable tool for engineers, particularly for preliminary design stages. However, they are limited to a two-dimensional analysis and are based on strong assumptions about the geometry of the structure, the spatial distribution and the behavior of the ground, as well as the excavation methods. Engineers are therefore increasingly using numerical modelling. They use computer codes to solve systems of differential equations using the finite element method or the finite difference method. Numerical modelling makes it possible to take into account more complex situations that are closer to reality, such as non-circular geometry of the structure, non-linear and delayed behavior of the rock mass, presence of heterogeneities and discontinuities, hydraulic regime for tunnels excavated below the water table, and phasing of the excavation process with its various support systems.

In many cases the numerical analysis is restricted to a two-dimensional model with the use of a deconfinement ratio to simulate the excavation phases. But in more complex situations, such as for conventional excavation with multiple stages, for which the assumptions of analytical methods and the restrictions of a two-dimensional numerical model are not valid anymore, three-dimensional modelling must be used. Three-dimensional modelling is also useful for a better assessment of the deconfinement rate taking into account an appropriate constitutive law for the ground as well as precise description of the support and pre-support conditions. However, it should be noted that, very often, a three-dimensional model can only simulate very local conditions, as for example in the case of the Saint-Martin-la-Porte access adit (Tran-Manh et al. 2015).

For a TBM excavation, two-dimensional numerical modelling may be appropriate since the section is circular, the excavation is carried out in full section and this excavation mode is most often used in homogeneous ground. However, only three-dimensional modelling can take into account the entire excavation process, the presence of an overcut, the conicity of the skirt, the installation of segmental lining and the injection of the filling material.

© The Author(s), under exclusive license to Springer Nature Switzerland AG 2022     141
M. Panet and J. Sulem, *Convergence-Confinement Method for Tunnel Design*,
Springer Tracts in Civil Engineering, https://doi.org/10.1007/978-3-030-93193-3_7

The finite element method, introduced in the 1950s, became very popular from the 1970s onwards, especially for solving structural mechanics and geotechnical problems (Zienkiewicz 1989, 1991a, b). From a variational formulation of the partial differential equations of the mechanical problem considered, an approximate solution is sought by discretizing the domain into contiguous elements of simple shape (polygons or polyhedra). The finite element mesh of the real domain is used to define the approximation space of the solution. The various fields studied (displacement, stress, pore pressure, temperature, etc.) are determined within each finite element from the values at the nodes of the element using simple linear or quadratic interpolation functions; continuity conditions are ensured at the nodes and boundary conditions are applied at the nodes located on the boundaries of the domain. By applying the variational principles of mechanics, the finite element method leads to the solution of a linear system written in matrix form:

$$KU + F_0 = F \tag{7.1}$$

where:

$F$    is the matrix of nodal forces;
$F_0$    is the matrix of nodal forces corresponding to the initial state of stress;
$K$    is the stiffness matrix;
$U$    is the matrix of nodal displacements.

The current power of computers makes it possible to study complex structures with a very large number of degrees of freedom. Numerous finite element computation codes suitable for geotechnical problems and relatively easy to use are available for engineers. However, these codes should not be used without a thorough understanding of how they work and what are their limitations. Checking of the convergence and of the mesh-size independence of the solution constitute essential stages of the numerical study, in order to avoid the production of erroneous results. In particular, constitutive laws exhibiting a strongly non-associated plastic behavior or a softening behavior of the material can lead to specific difficulties during their implementation in finite element codes. Softening behavior favors strain localization in the form of shear bands. Understanding and modeling strain localization phenomena has proven to be very useful for understanding failure mechanisms (Rudnicki and Rice 1975; Vardoulakis and Sulem 1995). The process of strain localization is seen as a material instability which can be predicted from the constitutive law of the material in the pre-failure regime. The conditions for the onset of shear banding are thus established by looking for the critical conditions for which the equations which describe the behavior of the material (in the pre-localization phase) can allow the existence of a bifurcation equilibrium state for which the deformation mode is localized into a planar band. In this approach, the initiation of failure in the form of an incipient shear band is described as a constitutive instability. From a mathematical point of view, this state of strain localization corresponds to a loss of uniqueness of the solution of the equations which govern the evolution of the mechanical system considered. The study of the system in the post-localization regime requires the implementation of specific mathematical methods in order to regularize the equations of the problem. These regularization methods are not a simple mathematical tool but are based on the modeling of the physical mechanisms that control this localization at the scale of the microstructure of the material. They resort to the theory of generalized continua which makes it possible to take into account the characteristics of the microstructure of materials in the formulation of constitutive laws (Germain 1973; Vardoulakis and Sulem 1995). Without the introduction of this regularization of the mathematical problem, the use of classical finite element codes becomes inappropriate in the softening regime of material behavior. This deficiency is manifested by numerical results which depend on the size of the mesh because the strains tend localize in a single element.

The reader will find in the recent paper of Pardoen and Collin (2017) an example of a robust hydro-mechanical finite element modeling for the computation of the deformations around a gallery of the underground laboratory of Meuse/Haute-Marne of Andra, taking into account the progressive development of localized deformation zones.

## 7.2 Numerical Modeling of Conventional Tunneling

For tunneling in difficult conditions, such as in squeezing ground, which requires several phases of excavation and support, numerical modelling makes it possible to explicitly simulate these different phases as well as the reprofiling phases of the sections. The size of the numerical model must be sufficiently large compared to the size of the structure (about 20 times the excavation opening) and the mesh is refined in the vicinity of the tunnel wall in order to correctly describe the strong stress and deformation gradients. The mesh must include the various temporary and permanent support structures that will be activated following the phasing of the excavation process. Different types of elements can be introduced to model the different support structures (sliding ribs, compressible blocks, concrete lining...). Interface elements are also introduced between the rock mass and the support by imposing different contact conditions depending on the nature of this contact (tied contact or relative sliding). The excavation is simulated by removing the elements of the ground at the face and by activating the support elements at a given distance from the face following the excavation history. An example is presented in Liu's (2020) doctoral thesis for the case of one of the Saint-Martin-la-Porte access galleries.

Sequential excavation is governed by the excavation step $s$ and the unsupported distance $d$. To simulate a continuous excavation, this step must be small enough (of the order of 0.4 times the tunnel diameter). To guarantee the quasi-static equilibrium of the structure, the time step must also be sufficiently small, and for rock mass with viscous behavior, this time step must be sufficiently small compared to the characteristic creep time of the rock.

The presence of a bolting at the face and/or the wall of the tunnel is simulated by introducing special structural elements or sometimes more simply by imposing a given pressure on the face and/or the wall.

Considering the complexity of the rock mass systems, the limited amount of information and the geological uncertainties, a major difficulty lies in the choice of the parameters to be included in the numerical models. Moreover, the short-term mechanical parameters as well as the time-dependent properties of the rock-mass can be significantly affected by the excavation method (De la Fuente et al. 2020) and the development of damaged and/or fractured zones around the tunnel.

## 7.3   Numerical Modeling of TBM Tunneling

The reader will find a detailed description of recent advances in finite element modeling of mechanized excavation in the articles by Kasper and Meschke (2004), and Zhao et al. (2012).

The numerical simulation of the excavation process with a tunnel boring machine (TBM) requires taking into account the complex interaction between the rock mass, the TBM with all its components and the support system (segmental lining, filling materials of the annular gap). This requires the use of three-dimensional models, which can be simplified to axisymmetric models in the case of an isotropic rock mass and in situ stresses.

For constant conditions in the tunneling direction, an elegant solution was proposed by Corbetta (1990), followed by Anagnostou (2007) which requires a single computational step (i.e. without an integration in the time-domain) to calculate the various mechanical and hydraulic fields. This approach considers that for an observer located on the tunnel face and moving with it, the stress, pore pressure and deformation fields are invariant in time. While the classical approach requires a step-by-step calculation simulating the excavation by the deactivation of elements of the rock mass and the installation of a support by the activation of elements of the lining, this method called 'stationary method' integrates in a single computational step all the progression of the excavation. This allows, using an appropriate change of variable $x^* = x - x_f(t)$ where $x$ is the position of the considered point and $x_f(t)$ the position of the face at time $t$, to perform faster computations by limiting the mesh refinement near the face and by avoiding the necessary phases of adaptive remeshing of the classical approach. This method is adapted to the case of a constant front speed and to repeatable excavation sequences, hence its interest in simulating a TBM excavation which is a continuous process.

The TBM is represented by a cylinder (rigid or deformable) of variable thickness to take into account its conicity. Shell elements can be used to simulate the segmental lining, interface elements or solid elements are placed between the segments and the rock to simulate the compressible filling material. Current computer codes allow for large deformation modeling. The shield and the cutting tools are simulated by plate elements with mechanical properties of steel. The cutting mechanism is complex and is usually represented in a simplified way by applying a pressure on the tunnel face (Berthoz et al. 2020).

The numerical model must be adapted to the type of TBM (e.g. single or double shielded TBM). The numerical modelling of the interaction between the shield and the rock mass and between the filling material and the concrete lining segments requires the introduction of interface elements of zero thickness. One difficulty is the appropriate choice of the sliding resistance at the interface for these elements in order to simulate different contact conditions (tied contact or sliding).

## 7.4 2D Numerical Modeling for the Application of the Convergence-Confinement Method

### 7.4.1 Principles of the Numerical Simulation

For the application of the convergence-confinement method in a two-dimensional numerical model, the computation consists of four phases:

- Application of the initial equilibrium state; this results in nodal forces on the intrados of the tunnel to be excavated.
- Simulation of the excavation before the installation of the lining; the nodal forces exerted at the intrados of the tunnel are decreased incrementally; during this phase, the lining elements are deactivated, i.e. their deformation modulus is taken equal to 0.
- Simulation of the installation of the lining by activation of the corresponding elements and of the end of the tunnel excavation by cancelling the nodal forces exerted at the intrados of the tunnel; this equilibrium state corresponds to the short-term conditions.
- Evaluation of the long term equilibrium conditions taking into account the time-dependent behavior of the rock mass and the long term mechanical characteristics of the lining.

### 7.4.2 Example

We present here an example of a numerical simulation performed with the finite difference code FLAC3D (Itasca 2011). In a ground of poor characteristics, a gallery with a circular section of radius $R = 3.15$ m is excavated at 600 m depth. The initial state of stress is assumed to be isotropic and equal to 16.2 MPa A shotcrete lining of thickness $e = 20$ cm is installed at a distance $d = 3.25$ m from the face.

The behavior of the ground is simulated by a elastic perfectly plastic Mohr–Coulomb model with the following mechanical characteristics:

Young's modulus: $E = 1600$ MPa

Poisson's ratio: $\nu = 0.3$

Cohesion: $C = 3.6$ MPa

Friction angle: $\phi = 26°$

Dilatancy angle: $\psi = 0$

The support is assumed to be elastic and its mechanical characteristics are:

Young's modulus: $E_s = 10000$ MPa

Poisson's ratio: $\nu_s = 0.25$

The 2D simulation requires the evaluation of the deconfinement ratio $\lambda_d$ at the support installation distance. For this purpose, an empirical formulation of the longitudinal displacement profile (LDP) can be used as described in Chap. 5. However, it is preferable, given the availability of numerical tools today, to perform a 3D axisymmetric calculation of the unsupported gallery to determine the LDP curve more accurately.

Figure 7.1 shows the mesh of the support and of the ground in the vicinity of the tunnel wall.

Figures 7.2 and 7.3 show the displacement and minor principal stress fields at final equilibrium, respectively (compressions are taken negative in FLAC3D code).

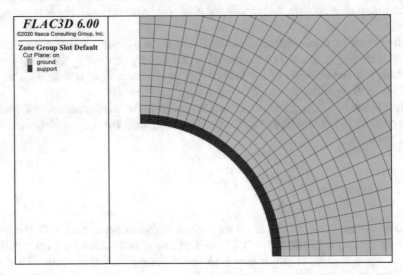

**Fig. 7.1** 2D numerical simulation—mesh of the support and of the ground in the vicinity of the tunnel wall

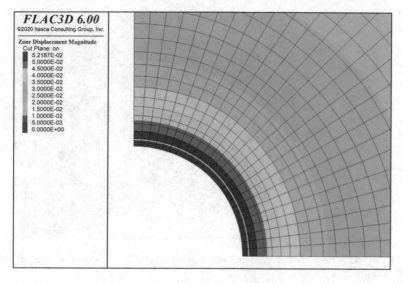

**Fig. 7.2**  2D numerical simulation—displacement magnitude

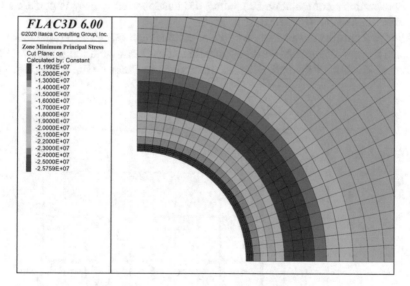

**Fig. 7.3**  2D numerical simulation—minimum principal stress

## 7.5  Example of 3D Numerical Simulation

The above example is now analyzed with a 3D numerical simulation. The three-dimensional axisymmetric mesh of the ground and of the support in a zone near the face is shown in Fig. 7.4. The excavation is performed with steps of 0.25 m.

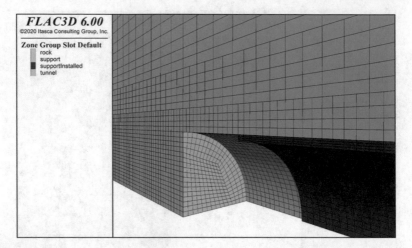

**Fig. 7.4.** 3D numerical simulation—mesh of the ground and of the support in the vicinity of the tunnel face

A preliminary computation simulating the unsupported gallery is performed to determinate the LDP (Fig. 7.5).

The LDP permits the evaluation of the deconfinement ratio $\lambda$ as a function of the distance from the front. This can be evaluated at each computation point using the analytical solution of the radial displacement given in Chap. 4 (Eq. 4.33) (Fig. 7.6).

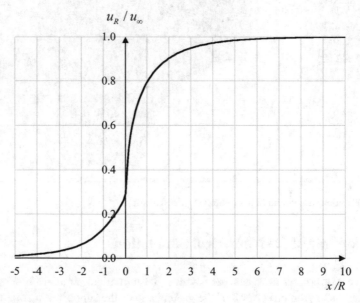

**Fig. 7.5** Longitudinal displacement profile (LDP) of the unsupported tunnel obtained by axisymmetric numerical simulation

**Fig. 7.6** Deconfinement ratio λ as a function of the distance from the face

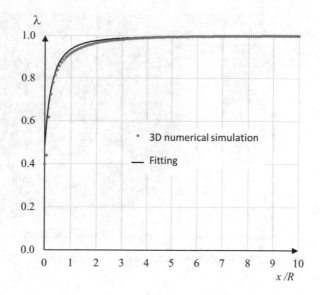

The computed values of the deconfinement ratio can be fitted using the empirical expression:

$$\lambda(x) = a_0 + (1 - a_0)\left(1 - \left[\frac{mR}{mR + x}\right]^2\right) \tag{7.2}$$

with $a_0 = 0.48$ et $m = 0.53$

Figures 7.7 and 7.8 show the displacement and minor principal stress fields, respectively (compressions are taken negative).

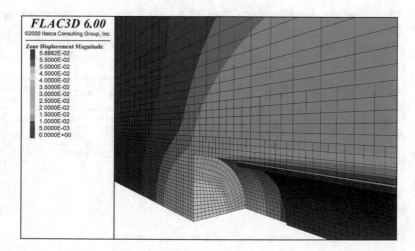

**Fig. 7.7** 3D numerical simulation—displacement magnitude

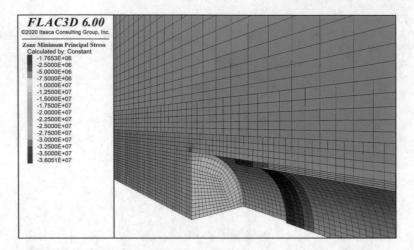

**Fig. 7.8** 3D numerical simulation—minor principal stress

It is interesting to compare the results obtained from this 3D numerical simulation with those obtained from the 2D numerical simulation and those obtained from the analytical approaches. This comparison is presented in Table 7.1 for different distances of support installation ($d = 1$ m, $d = 3.25$ m, $d = 2R = 6.25$ m).

These results show that the classical 2D analytical approach gives a very good evaluation of the equilibrium radial displacement $u_{eq}$ at final equilibrium but underestimates the support pressure $p_s$ and thus the normal force $\sigma_N$ in the support. The implicit method of Nguyen-Minh and Guo (see Sect. 5.2.2) gives a good evaluation of the support pressure at equilibrium, especially when the support is placed near the face. The displacement is underestimated but the accuracy remains acceptable.

**Table 7.1** Comparison of results obtained from 3D numerical simulation and from analytical methods

| | 3D numerical results | | | 2D analytical results Implicit method of Nguyen-Minh and Guo | | | 2D analytical result Classical approach | | |
|---|---|---|---|---|---|---|---|---|---|
| $d$ (m) | $u_{eq}$ (m) | $p_s$ (MPa) | $\sigma_N$ (MPa) | $u_{eq}$ (m) | $p_s$ (MPa) | $\sigma_N$ (MPa) | $u_{eq}$ (m) | $p_s$ (MPa) | $\sigma_N$ (MPa) |
| 1 | 0.044 | 2.31 | 36.39 | 0.041 | 2.37 | 37.33 | 0.044 | 1.82 | 28.67 |
| 3.25 | 0.056 | 1.07 | 16.85 | 0.052 | 0.93 | 14.65 | 0.054 | 0.66 | 10.40 |
| 6.25 | 0.059 | 0.50 | 7.81 | 0.058 | 0.38 | 5.92 | 0.059 | 0.26 | 4.08 |

# References

Anagnostou G (2007) The one-step solution of the advancing tunnel heading problem. In: ECCOMAS thematic conference on computational methods in tunnelling, p 17

Berthoz N, Branque D, Subrin D (2020) Déplacements induits par les tunneliers: rétro-analyse de chantiers en milieu urbain sur la base de calculs éléments finis en section courante. Rev Fr Géotech 164:1

Corbetta F (1990) Nouvelles methodes d'études des tunnels profonds—Calculs analytiques et numériques. Thèse de doctorat de l'Ecole Nationale Supérieure des Mines de Paris

De la Fuente M, Sulem J, Taherzadeh R, Subrin D (2020) Tunneling in squeezing ground : effect of the excavation method. Rock Mech Rock Eng 53:601–623

Germain P (1973) The method of virtual power in continuum mechanics. Part 2 : microstructure. SIAM J Appl Math 25(3):556–575

ITASCA (2011) Fast Lagrangian analysis of continua (FLAC3D). Itasca Consulting Group Inc., Minnesota

Kasper T, Meschke G (2004) A 3D finite element simulation model for TBM tunnelling in soft ground. Int J Numer Anal Meth Geomech 28(14):1441–1460

Liu Y (2020) Modélisation du comportement différé et anisotrope des terrains fortement tectonisés: Application au creusement d'une galerie de reconnaissance du tunnel de base de la liaison Lyon-Turin. Thèse de doctorat de l'Université Paris-Est

Pardoen B, Collin F (2017) Modelling the influence of strain localisation and viscosity on the behaviour of underground drifts drilled in claystone. Comput Geotech 85:351–367

Rudnicki JW, Rice JR (1975) Conditions for the localization of deformation in pressure-sensitive dilatant materials. J Mech Phys Solids 23:371–394

Tran-Manh H, Sulem J, Subrin D, Billaux D (2015) Anisotropic time-dependent modeling of tunnel excavation in squeezing ground. Rock Mech Rock Eng 48(6):2301–2317

Vardoulakis I, Sulem J (1995) Bifurcation analysis in geomechanics. CRC Press

Zhao K, Janutolo M, Barla G (2012) A completely 3D model for the simulation of mechanized tunnel excavation. Rock Mech Rock Eng 45(4):475–497

Zienkiewicz OC (1989) The finite element method in structural and continuum mechanics. Mc Graw Hill, London

Zienkiewicz OC, Taylor RL (1991) The finite element method. In: Basic formulation for linear problems, vol 1. Mc Graw Hill, London

Zienkiewicz OC, Taylor RL (1991) The finite element method. In: Solid and fluid mechanics, dynamics and non linearity, vol 2. Mc Graw Hill, London

Printed in the United States
by Baker & Taylor Publisher Services